AF229436

Agathe Keller

Expounding the Mathematical Seed

Volume 1: The Translation

A Translation of Bhāskara I on the Mathematical Chapter
of the Āryabhatīya

Birkhäuser Verlag
Basel · Boston · Berlin

Author

Agathe Keller
Rehseis CNRS
Centre Javelot
2 place Jussieu
75251 Paris Cedex 05
e-mail: kellera@paris7.jussieu.fr

A CIP catalogue record for this book is available from the Library of Congress, Washington D.C., USA

Bibliographic information published by Die Deutsche Bibliothek
Die Deutsche Bibliothek lists this publication in the Deutsche Nationalbiographie; detailed bibliographic data is available in the internet at http://dnb.ddb.de

ISBN 3-7643-7291-5 Birkhäuser Verlag, Basel – Boston – Berlin

© 2006 Birkhäuser Verlag, P.O.Box 133, CH-4010 Basel, Switzerland
Part of Springer Science+Business Media
Cover design: Micha Lotrovsky, CH-4106 Therwil, Switzerland
Cover illustration: The cover illustration is a free representation of
"Rat and Hawk" (made by Mukesh)
Printed on acid-free paper produced from chlorine-free pulp. TCF ∞
Printed in Germany

Vol. 1/SN 30:	ISBN 10: 3-7643-7291-5	e-ISBN: 3-7643-7592-2
	ISBN 13: 978-3-7643-7291-0	
Vol. 2/SN 31:	ISBN 10: 3-7643-7292-3	e-ISBN: 3-7643-7593-0
	ISBN 13: 978-3-7643-7292-7	
Set SN 30/31:	ISBN 10: 3-7643-7299-0	e-ISBN: 3-7643-7594-9
	ISBN 13: 978-3-7643-7299-6	

9 8 7 6 5 4 3 2 1

Contents

Acknowledgments

This book would not have been possible without the tireless endeavors of K. Chemla and T. Hayashi. They have followed my work from its hesitating beginnings until the very last details of its final form. Many of the reflections, the fine precise forms of the translation are due to them. J. Bronkhorst has generously reread and discussed the translation, giving me the impulse to achieve a manuscript fit for printing. It would now be difficult to isolate all that the final form owes to these three people. However, I bear all the responsibilities for the mistakes printed here. My study of Bhāskara's commentary was funded by grants from the Chancellerie des Universits de Paris, the Indian's goverment's ICCR, the French MAE, the Japanese Monbusho, and the Fyssen Foundation. Computor equipment was generously provided by KEPLER, sarl. Finally, the loving support of my family and friends has given me the strength to start, continue and accomplish this work.

Abreviations and Symbols

When referring to parts of the treatise, the *Āryabhaṭīya*, we will use the abbreviation: "Ab". A first number will indicate the chapter referred to, and a second the verse number; the letters "abcd" refer to each quarter of the verse. For example, "Ab. 2. 6. cd" means the two last quarters of verse 6 in the second chapter of the *Āryabhaṭīya*.

With the same numbering system, BAB refers to Bhāskara's commentary. Mbh and Lbh, refer respectively to the *Māhabhāskarīya* and the *Laghubhāskarīya*, two treatises written by the commentator, Bhāskara.

[] refers to the editor's additions;

⟨⟩ indicates the translator's additions;

() provides elements given for the sake of clarity. This includes the transliteration of Sanskrit words.

Introduction

This book presents an English translation of a VIIth century Sanskrit commentary written by an astronomer called Bhāskara. He is often referred to as Bhāskara I or "the elder Bhāskara" to distinguish him from a XIIth century astronomer of the Indian subcontinent bearing the same name, Bhāskara II or "the younger Bhāskara".

In this commentary, Bhāskara I glosses a Vth century versified astronomical treatise, the *Āryabhaṭīya* of Āryabhaṭa. The *Āryabhaṭīya* has four chapters, the second concentrates on *gaṇita* or mathematics. This book is a translation of Bhāskara I's commentary on the mathematical chapter of the *Āryabhaṭīya*. It is based on the edition of the text made by K. S. Shukla for the Indian National Science Academy (INSA) in 1976[1].

How to read this book?

This work is in two volumes. Volume I contains an Introduction and the literal translation. Because Bhāskara's text alone is difficult to understand, I have added for each verse's commentary a supplement which discusses the linguistic and mathematical matter exposed by the commentator. These supplements are gathered in volume II, which also contains glossaries and the bibliography. The two volumes should be read simultaneously.

This Introduction aims at providing a general background for the translation. I would like to help the reader with some of the technical difficulties of the commentary, in appearance barren and rebutting. My ambition goes also beyond this point: I think reading Bhāskara can become a stimulating and pleasant experience altogether.

The Introduction is divided into three sections. The first places Bhāskara's text within its historic context, the second looks at its mathematical contents, the third analyzes the relations between the commentary and the treatise.

---

[1]The Bibliography is at the end of volume II, on p.227. The edition is listed under [Shukla 1976]. The conventions used for the translation are listed in the next section on p.1.

Let us start by describing Bhāskara's commentary: we will shortly observe where it stands within the history of mathematics and astronomy in India and specify, afterwards, the type of mathematical text Bhāskara has written.

A   Situating Bhāskara's commentary

1   A brief historical account

The following is a short sketch of the position of Bhāskara's *Āryabhaṭīyabhāṣya* ("commentary of the *Āryabhaṭīya*", the title of his book) among the known texts of the history of mathematics and astronomy in India[2].

The oldest mathematical and astronomical corpus that has been handed down to us in this geographical area are related to the vedas.

The vedas are a set of religious poems. They are the oldest known texts of Indian culture. These poems also form the basis on which, later, Hinduism developed. This is probably why their date and origin are still subject to intense historical and philological debates[3]. These poems have been commented upon in all sorts of ways, grammatically, philosophically, religiously, ritually. The sum of these commentaries is called the *vedāṅgas*. There is a mathematical component to these texts which is related to the construction of the altars used in religious sacrifices. These mathematical writings are called the *śulbasūtras*[4]. They are of composite nature, have different authors and thus several dates. The earliest is generally considered to be the *Baudhyānaśulbasūtra* of circa 600 B.C.[5] The *śulbasūtras* typically describe constructions with layered bricks or the delimitation of areas with ropes. They are not, however, devoid of testimonies of general mathematical reflections. For instance, they state the "Pythagoras Theorem".[6] Reading Bhāskara's commentary, one comes across objects and features (such as strings) that are inherited from this tradition[7].

Bhāskara's text, however, belongs to a different mathematical tradition. Indeed, the *śulbasūtras*, together with the *vedāṅgajyotiṣa* (circa. 200 B. C.), an astronomical treatise with no mathematical content, are historically followed by a gap:

---

[2]For more detailed accounts one may refer to [Pingree 1981], [Datta and Singh 1980], [Bag 1976]. Many books have been published in India on this subject, they usually recollect what was printed in the afore mentioned classics.

[3]The nature and scope of these debates have been analyzed in [Bryant 2001]. While most Indologists will agree to ascribe to the vedas the date of ca. 1500 B. C., traditional pandits and scholars with a bend towards hindu nationalism might quote very old dates, starting with 4000 or 5000 B. C. and going further back.

[4]All the edited and translated texts are gathered in [Bag & Sen 1983].

[5][CESS, volume 4].

[6]For more details see [Sarasvati 1979], [Bag 1976], [Datta & Singh 1980] and [Hayashi 1994, p. 118].

[7]The question of the posterity of the *śulbasūtras* in Bhāskara's commentary remains an area open for further investigation.

after that, the Hindu tradition[8] has not handed down to us any mathematical or astronomical text dated before the Vth century A.D.

At that time, two synthetic treatises come to light: the *Pañcasiddhānta* of Varāhamihira[9] and the *Āryabhaṭīya* of Āryabhaṭa[10]. The importance of the *Āryabhaṭīya* for the subsequent astronomical reflection in the Indian subcontinent can be measured by the number of commentaries it gave rise to and the controversies it sparked. Indeed no less than 18 commentaries have been recorded on the *Āryabhaṭīya*, some written as late as the end of the XIXth century[11]. Both texts are typical Sanskrit treatises: they are written in short concise verses. They are compendiums. Varāhamihira's composition, for instance, as indicated by its title which means "the five *siddhāntas*", summarizes five treatises[12].

Bhāskara, probably a marathi astronomer[13], has written the oldest commentary of the *Āryabhaṭīya* that has been handed down to us. Consequently, it is the oldest known Sanskrit prose text in astronomy and mathematics. According to his own testimony it was composed in 629 A.D.

Very little is known about Bhāskara and his life, only that he is also the author of two astronomical treatises in the line of Āryabhaṭa's school, the *Māhabhāskarīya* and the *Laghubhāskarīya*[14].

Other VIIth century mathematical and astronomical texts in Sanskrit have been handed down to us. Brahmagupta's treatise the *Brahmasphuṭasiddhānta*[15] would have been written in 628 A.D. and, in the latest critical asssessment of its datations, the *Bhakhshālī Manuscript*[16], a fragmentary prose, is also roughly ascribed to the VIIth century[17].

Thus, the VIIth century appears as the first blossoming of a renewed mathematical and astronomical tradition. Thereafter, a continuous flow of treatises and commen-

---

[8]The Hindu tradition is the sum of mathematical works developed by Hindu authors. It includes almost all the texts written in Sanskrit, although Hindu authors have also written in other dialects. Buddhists, for which we do not have any early testimony of mathematical writings, and Jain authors almost systematically wrote in their own dialects. At the beginning of the VIth century a council in Valabhī, a town mentioned in Bhāskara's examples, fixed the Jain canon which includes astronomical and mathematical texts. Although not written in Sanskrit, some quotations of these works are found in our commentary. These Jain texts testify to the existence of mathematical and astronomical knowledge developed outside of the Hindu tradition, prior to the VIIth century, and probably before the Vth century as well.

[9][Neugebauer & Pingree 1971].

[10][Sharma & Shukla 1976].

[11][Sharma & Shukla 1976; xxxv-lviii].

[12]Concerning the name *siddhānta* for astronomical treatises, see [Pingree 1981].

[13][Shukla 1976; xxv-xxx], [CESS; volume 4, p. 297].

[14]These texts have been edited and translated by K. S. Shukla: [Shukla 1960], [Shukla 1963]. They had also been previously edited with commentaries, see [Apaṭe 1946] and [Sastri 1957]. For more details one can refer to the entry Bhāskara in the [CESS, volume 4, p. 297-299; volume 5].

[15][Dvivedi 1902].

[16][Hayashi 1995].

[17]For a discussion of the time when the text would have been written, see [Hayashi 1995, p. 148-149].

taries in Sanskrit were produced and preserved. At that time, Sanskrit astronomical texts and knowledge spread outside the frontiers of the Indian subcontinent: by the IXth century, there where Indian astronomers at the Tang courts and most probably the first Indian treatises were translated into Arabic. The precise story of these astronomical and mathematical creations still needs to be written. However, they provide testimony to the florescence of these disciplines in India during this period. By the XVIIth century, in turn, texts in Arabic and Persian started to imprint their mark on the astronomical knowledge of India, announcing a new way of practicing this discipline.

Bhāskara's prose writing is therefore important because it provides information on the beginning of one of the richest moments in the development of mathematics and astronomy in ancient India. It can, indeed, furnish clues to the relations these mathematics have to the former tradition of Vedic geometry. Furthermore, Bhāskara's *Āryabhaṭīyabhāṣya* proposes an interpretation of an important Vth century treatise. We will see later that it is certainly *his reading* of this text. Furthermore, Bhāskara's commentary does not only shed a light on the treatise, it also provides detailed insights on the authors' own mathematical and astronomical practices.

2   Text, Edition and Manuscript

Bhāskara's mathematics are not unknown to historians of mathematics. An edition of his commentary was published in 1976 by K. S. Shukla for the Indian National Science Academy[18], the completion of a series that had started at the University of Lucknow in the 1960's with the publication of editions and translations of Bhāskara's two other astronomical treatises[19]. These were followed by a number of articles by the same author on Bhāskara's mathematics[20]. Books published in India will often refer to him for his contributions to the pulverizer and his arithmetics, if not for his trigonometry or his use of irrational numbers. Bhāskara is indeed famous and glorious, but nothing much is usually said beyond broad generalities. Among the reasons that could be ascribed to such an attitude, one should insist on the difficulties presented by the edited text itself. It is difficult to read.

2.1   On the edition and its manuscripts

This difficulty can be ascribed to the scarcity and state of the sources that were used while elaborating the edition.

---

[18][Shukla 1976].
[19][Shukla 1960] and [Shukla 1962].
[20][Shukla 1971 a], [Shukla 1971 b], [Shukla 1972 a], and [Shukla 1972 b].

Indeed, only six manuscripts[21] of the commentary are known to us. Five of them were used to elaborate Shukla's edition. These five belong to the Kerala University Oriental Manuscripts Library (KUOML) in Trivandrum and one belongs to the Indian Office in London[22]. All the manuscripts used in the edition prepared by K. S. Shukla have the same source. This means they all have the same basic pattern of mistakes, each version having its own additional ones as well. They are all incomplete. Shukla's edition of the text has used a later commentary on the text inspired by Bhāskara's commentary, to provide a gloss of the end of the last chapter of the treatise. The fact that this edition relies on a single faulty source is probably one of the reasons why Bhāskara's commentary at times seems obscure or nonsensical. As many old Indian manuscripts still belong to private families or remain hidden in ill-classified libraries, one can still hope to find supplementary recensions that would enable a revision the edition.

While the lack of primary material is a major difficulty, other problems arise from the quality of the edition itself. K. S. Shukla has indeed performed the tedious meticulous work required for an edition. However some aspects of this endeavor, retrospectively, raise some questions. Let us first note that no dating of the manuscripts or attempt to trace their history has been taken up. Secondly, nonsensical or problematic parts have not been systematically pointed out and discussed. K. S. Shukla has indeed provided in many cases alternative readings. However, these are never justified and sometimes go contrary to the sensible manuscript readings that he gives in footnotes[23]. But in many cases, nonsensical sentences are found in the text without any comment at all. A third problem arises as editorial choices concerning textual arrangements (such as diagrams and number dispositions) are often, if not systematically, implicit. I have consulted four of the six manuscripts of the text and can testify that dispositions of numbers and diagrams vary from manuscript to manuscript. Discrepancies between the printed text and the manuscript further deepen the already existing gap between the written text and the manuscripts themselves. Concerning the latter, manuscripts and edition are separated by more than 1000 years of mathematical practices[24]. Consequently, all study of diagrams, or of *bindus* as representing zero should be carried out carefully.

---

[21] Five of which are made of dried and treated palm leaves which were carved and then inked, a traditional technique in the Indian subcontinent. Palm-leaf manuscripts do not keep well, and thus Sanskrit texts have generally been preserved in a greater number on paper manuscripts.

[22] Shukla has used four from the KUOML and the one from the BO. A fifth manuscript was uncovered by D. Pingree at the KUOML. As one of the manuscripts of the KUOML is presently lost it is difficult to know if the "new one" is the misplaced old one or not. Furthermore, this manuscript is so dark that its contents cannot be retrieved anymore.

[23] As specified in the next section, p. 1, when this was the case, the translation adopted was that of the manuscript readings.

[24] A case study on the dispositions of the Rule of Three has been studied in [Sarma 2002] which underlines such discrepancies. Palmleaf manuscripts can not be much older than 500 years.

2.2 Treatise versus commentary

Bhāskara's fame is also obliterated by Āryabhaṭa's celebrity. Āryabhaṭa is a figure that all primary educated Indians know. He is celebrated as India's first astronomer. India's first satellite was named after him. This reputation rests upon the understanding we have of his works and achievements. As we will attempt to show later on, for this we need to rely on his commentators. And indeed, historically, many who achieved understanding of Āryabhaṭa have been indebted to Bhāskara. The first publication of Āryabhaṭa's text in Sanskrit[25] was accompanied by a commentary by Parameśvara, another astronomer and commentator on Āryabhaṭa. Parameśvara knew Bhāskara's commentary and relied on it. Subsequent translations in English and German have first relied on Parameśvara's commentary and, when it came to be known, on Bhāskara's commentary as well[26]. Bhāskara's importance can be measured by looking at the different understandings scholars (traditional and contemporary) have had of Āryabhaṭa's text. T. Hayashi has shown how Bhāskara's misreading of verse 12 of the mathematical chapter of the *Āryabhaṭīya*[27] has induced a long chain of misleading interpretations[28].

Why then, has the commentator been "swept under the rug", to use a French expression? The bias, privileging the treatise over its commentaries has partly its origin in the field of Indology itself. Indeed, even if we do not restrict ourselves to the astronomical and mathematical texts, the great bulk of Sanskrit scholarly literature is commentarial. Moreover, in India, commentaries could be as important as the treatises they glossed. For example, for the grammatical tradition, the *Mahābhāṣya* is probably as important as the text it comments, the *Aṣṭādhyāyi* of Pāṇini. However, despite their importance, there exists almost no thorough study on the genre of Sanskrit commentaries produced in a discipline whose object is after all ancient Indian texts[29]. A similar disregard of commentaries can also be found in the field of history of mathematics. Thus Reviel Netz's study of late medieval Euclidean commentaries, in an attempt to rehabilitate their importance, is not devoid of such prejudices[30]. The disregard of commentaries in both disciplines is probably a contemporary remnant of the Renaissance disregard for this kind of literature, a hint that these fields of scholarship were born in Europe. Whatever the reason, the consequence has been that the contents of Bhāskara's astronomical and mathematical texts has little been detailed in secondary literature.

Our aim is thus to focus on Bhāskara's work, highlighting two aspects: his interpretations of Āryabhaṭa's verses and his personal mathematical input. Let us

---

[25][Kern 1874].

[26]See [Sengupta 1927], [Clark 1930], and [Sharma & Shukla 1976].

[27]From now on, all verses referred to belong to the mathematical chapter of the *Āryabhaṭīya*, unless otherwise stated.

[28]See [Hayashi 1997a].

[29]Let us nevertheless mention [Renou 1963], [Bronkhorst 1990], [Bronkhorst 1991], [Houben 1995] and [Filliozat 1988 b, Appendix], which are first attempts in specific disciplines, such as grammar, and at given times (Bronkhorst looks at the the VIIth century).

[30]See [Netz 1999] and as an answer [Chemla 2000], [Bernard 2003].

specify briefly what Āryabhaṭa's verses are and how the commentary is structured before giving an overview of its contents.

2.3 Āryabhaṭa's sūtras

The *Āryabhaṭīya* is composed of concise verses, mostly in the famously difficult *āryā* (the first chapter being an exception and being written in the *gītikā* verse[31]). These hermetic rules are known as *sūtras*. Āryabhaṭa's *sūtras* can be definitions (like verse 3^{32} which defines squares and cubes) or procedures (like verse 4^{33} which provides an algorithm to extract square roots). Some are a blend of such characterizations (thus verse 2^{34} defines the decimal place value notation and the process to note such numbers). They manipulate technical mathematical objects such as numbers, geometrical figures and equations. Āryabhaṭa's *sūtras* use puns, which gives to them an additional mnemonic flavor. Let us look, for instance, at Verse 4:

> One should divide, constantly, the non-square ⟨place⟩ by twice the square-root|
> When the square has been subtracted from the square ⟨place⟩, the quotient is the root in a different place‖
>
> *bhāgaṃ hared avargān nityaṃ dviguṇena vargamūlena|*
> *vargād varge śuddhe labdhaṃ sthānāntare mūlam‖*

As analyzed in the supplement on this verse and its commentary[35], the rule describes the core of an iterative process: the algorithm computes the square-root of a number noted with the decimal place value notation. It is concise in the sense that one needs to supply words to understand with more clarity what is referred to. This is indicated in the translation by triangular brackets (⟨⟩[36]). Its brevity is connected to a pun: one does not know if the "squares" referred to in the verse are square numbers or square places (a place corresponding to a pair/square power of ten in the decimal place value notation). Obviously, this pun has also a mathematical signification, providing a link between square places and square numbers. Even when Āryabhaṭa's verses do not handle such elaborate techniques, they often only state the core of a process. We often do not know what is required and what is sought, or what are all the different steps one should follow to complete the algorithm. Indeed, such rules call for a commentary.

---

[31] For more precision on the form of the treatise, one can refer to [Keller 2000; I] or see [Sharma & Shukla 1976].

[32] See BAB.2.3, volume I, p. 13-18.

[33] See BAB.2.4, volume I, p. 20.

[34] See BAB.2.2, volume I, p. 10.

[35] See volume II, p. 15.

[36] Conventions for such symbols are listed in volume I, p. ix.

2.4 Structure of the commentary

Bhāskara's commentary follows a systematical pattern. This structure can be found in other mathematical commentaries as well[37]. He glosses Āryabhaṭa's verses in due order.

The structure of each verse commentary is summarized in Table 1.

Table 1: Structure of a verse commentary

Introductory sentence
Quotation of the half, whole, one and a half or two verses to be commented
General commentary, e.g.
Word to word gloss, staged discussions, general explanations and verifications
"Solved examples" (*uddeśaka*)
Versified Problem
"Setting-down" (*nyāsa*)
"procedure" (*karaṇa*)

Each verse gloss starts by an introductory sentence which gives a summary of the subject treated in the verse. This introduction is followed by a quotation of the verse(s) to be commented. It is succeeded by what we have called a "general commentary". This portion of the text is a word to word gloss of the verse, where syntax ambiguities are lifted, words supplied and technical vocabulary justified and explained. This is also the part of the commentary which will present staged dialogs and discussions justifying Bhāskara's interpretation of Āryabhaṭa's rule. This "general commentary" is followed by a succession of solved examples. Each solved example once again is molded into a quite systematical structure. It is first announced as an *uddeśaka*. It is followed by a versified problem. The versified problem precedes a "setting-down" (*nyāsa*), where numbers are disposed, diagrams drawn as they will be used on a working surface from which the problem will be solved. This is followed by a resolution of the problem called *karaṇa* ("procedure").

Having thus described Bhāskara's commentary and located it historically, let us now turn to its contents.

In the following section we will present a structural overview of the mathematics of Bhāskara's commentary. A second section will attempt to draw the attention of the reader to the characteristics of the *Āryabhaṭīyabhāṣya* as a mathematical *commentary* of the Sanskrit tradition.

---

[37]See for instance [Jain 1995], [Patte].

B   The mathematical matter

The mathematical chapter of the *Āryabhaṭīya* contains a great variety of procedures, as summarized in Table 2 on page xx.

Subjects treated range from computing the volume of an equilateral tetrahedron (verse 6) to the interest on a loaned capital (verse 25), from computations on series (verses 19-22) to an elaborate process to solve a Diophantine equation (verse 32-33). All of these procedures are given in succession, without any structural comment. It is the commentator, Bhāskara, who introduces several ways to classify them[38]. We will take up one such classification that seems to contain a relevant thread to synthesize Āryabhaṭa's and Bhāskara's treatment of *gaṇita* (mathematics/computations [39]): namely the distinction between *rāśigaṇita* ("mathematics of quantities") and *kṣetragaṇita* ("mathematics of fields"[40]). Naturally, Bhāskara's "arithmetics" or "geometry" does not always distribute procedures into the categories we would expect them to be allotted to. For instance, rules on series are considered as part of geometry. Furthermore, these classifications are not exclusive and a procedure can bear both an "arithmetical" and a "geometrical" interpretation[41]. Let us insist here that we are considering Bhāskara's practice of mathematics as we know very little of Āryabhaṭa's mathematics.

We will follow the opposition between the categories of *rāśigaṇita* and *kṣetragaṇita* to list a certain number of characteristics of mathematics as practiced by Bhāskara. While doing so, we will underline the ambiguities and uncertainties that these subdivisions raise. Our stress will be on the *practices* of mathematics that Bhāskara's commentary testifies of. Having examined separately procedures belonging to "arithmetic" and to "geometry" in Bhāskara's sense, we will analyze what are the relations entertained by these two disciplines. We will then turn, to articulating the broader link of mathematics with astronomy.

1   Bhāskara's arithmetics

Let us first look at the quantities used by Bhāskara before examining some aspects of his arithmetical practices. These activities and objects belong to the commentary. Unless stated, they are not mentioned in the treatise.

---

[38]I have analyzed these classifications and the definition of *gaṇita* in [Keller forthcoming].

[39]This word is used to refer to the subject or field "mathematics" but can also name any computation. I have discussed this polysemy in [Keller 2000; volume 1, II. 1] and in [Keller forthcoming]. This is also briefly alluded to below, on p.xxxviii and in the Glossary at the end of volume II (p.197.)

[40]*Kṣetra*, "field", is the Sanskrit name for geometrical figures.

[41]This will be discussed in more detail below on p. xxxiv.

Table 2: Contents of the Chapter on mathematics (*gaṇitapāda*)

Verse 1	Prayer
Verse 2	Definition of the decimal place value notation
Verse 3	Geometrical and arithmetical definition of the square and the cube
Verse 4	Square root extraction
Verse 5	Cube root extraction
Verse 6	Area of the triangle, volume of an equilateral tetrahedron
Verse 7	Area of the circle, volume of the sphere
Verse 8	Area of a trapezium, length of inner segments
Verse 9	Area of all plane figures and chord subtending the sixth part of a circle
Verse 10	Approximate ratio in a circle, of a given diameter to its circumference
Verses 11-12	Derivation of sine and sine differences tables
Verse 13	Tools to construct circle, quadrilaterals and triangles, verticality and horizontality
Verses 14-16	Gnomons
Verse 17	Pythagoras Theorem and inner segments in a circle
Verse 18	Intersection of two circles
Verses 19-22	Series
Verses 23-24	Finding two quantities knowing their sum and squares or product and difference
Verse 25	Commercial Problem
Verse 26	Rule of Three
Verse 27	Computations with fractions
Verse 28	Inverting procedures
Verse 29	Series/First degree equation with several unknowns
Verse 30	First degree equation with one unknown
Verse 31	Time of meeting
Verses 32-33	Pulverizer (Indeterminate analysis)

1.1 Naming and noting numbers

There is a difference between the way one names a number with words, and the way it is noted, on a working surface, to be used in computations.

1.1.a Naming numbers Sanskrit uses diverse ways of naming numbers, Bhāskara resorts to many. There exist technical terms for numbers, which bear Indo-European characteristics: thus the name for digits are *eka, dva, tri, catur, pañca, ṣaḍ, sapta, aṣṭa, nava*. Some numbers can alternatively be named by operations of which they are the result, thus *ekonavaviṃśati* (twenty minus one) for nineteen or *trisapta* (three ⟨times⟩ seven) for twenty-one. Numbers, especially digits, can also be named by a metaphor which is indicative of a number. Thus, the moon (*śaśin*) refers to one. A pair of twin gods, the Aśvins, can name the number 2, etc. As the last example shows, most of these metaphors rest upon images that spring from India's rich mythological tradition. These metaphors are used essentially when giving very big integer numbers: the commentator then enumerates in a compound (*dvandva*) the digits that constitute the number when it is noted with the decimal place value system, by following the order of increasing values of power of tens[42]. This was probably a way to ensure that no mistake was made when the number was noted. All of these devices can also be used to give the value of a fraction.The variety and complexity with which numbers are named require a mathematical effort: they need to be translated into a form that enables them to be easily manipulated on a working surface. This probably explains why a rule is actually given explaining how to note numbers (Ab.2.2). A glossary of the names of numbers can be found in volume II[43].

1.1.b Decimal place value notation To write down numbers, Bhāskara uses the decimal place value notation that Āryabhaṭa defines in verse 2 of the chapter on mathematics. The commentator is well aware of the advantages that this notation has on other types of notations[44]. No procedures for elementary operations are given in the text[45]. However, the rules Bhāskara gives to square and cube higher numbers and Āryabhaṭa's procedures to extract square roots rest upon such a notation of numbers and uses its properties. It is therefore highly probable that the same held true for elementary operations. Units (*rūpa*) accumulated produce digits (*aṅka*) and numbers (*saṅkhyā*). The word *aṅka* means "sign" or "mark" and could therefore refer to the symbols used to note the digits rather than to their

---

[42]The digits are therefore enumerated in an order that is opposite to the one with which they are noted. All of this is discussed and detailed in [Keller 2000; I.2.2.1]. For specifications on how the numbers have been translated see the next section, p. 2.

[43]See volume II, p. 221.

[44]This can be inferred from the rather obscure opening paragraph of the commentary of verse 2. It raises questions as to whether another system for noting numbers was prevalent in India. See BAB.2.2, volume I, p.10.

[45]Later texts describe such operations, using the decimal place value notation in the algorithms.

value. However these distinctions aren't used systematically, the word *saṅkhyā* often refers to digits.

1.1.c Noting fractions   In the printed edition[46] of Bhāskara's commentary two ways to note rational numbers can be observed. Both forms are noted in a column. There are "fractions", as we are used to them, with a numerator and a denominator. The fraction $\frac{a}{b}$ is noted $\begin{smallmatrix}a\\b\end{smallmatrix}$, where a and b are noted, of course, with the decimal place value system. Moreover, rational numbers are manipulated in another form we call "fractionary numbers", consisting of an integer number plus or minus a fraction smaller then 1. In this case, $c - \frac{a}{b}$ is noted as $\begin{smallmatrix}c\\a°\\b\end{smallmatrix}$; and $c + \frac{a}{b}$ is noted similarly, omitting the circle next to the numerator, $\begin{smallmatrix}c\\a\\b\end{smallmatrix}$. The integer part of a fractionary quantity is called *uparirāśi* , the "quantity above". The fraction used within a fractionary number is usually called *aṃśa*. When the value of a fraction is given, this very word, which originally means "part", is suffixed either to the denominator or the numerator of the fraction[47]. It is also used as a technical term to name numerators of fractions. Denominators are then referred to as *cheda* which also means "part".

The commentary provides rules to transform fractionary numbers into fractions and vice-versa. In practice, the fractionary form of a number is clearly distinguished from that of a fraction. Fractions bigger than 1 seem to have been perceived by Bhāskara as temporary notations used while computing. They are used in intermediary steps of procedures and not as results. Although distinguished in practice, these two quantities do not have separate names. *Saccheda* ("with a denominator") can refer to both a fractionary number or a fraction. Similarly, the word *bhinna* (different, part) can name one or the other form. This ambiguity enables Bhāskara to provide a double reading of the rule given in the second half of verse 27 [48] : it can be seen as a rule to change fractionary numbers into fractions or as a rule to reduce two fractions to the same denominator.

Bhāskara seems to have thought of a rational number as an integer or a sequence of integers for which no refined measuring unit existed. This can be seen very clearly in the examples of the commentary on the Rule of Three: a list of different integers is obtained by successively refining measuring unit, the last number of the list being a fraction smaller than 1. Never is the value of a fraction bigger than 1 stated, even though fractions bigger than 1 are often noted[49].

---

[46]These dispositions can also be seen in the manuscripts we have consulted.

[47]This is also stated in the *aṃśa* entry of the Glossary, volume II, p. 197.

[48]See volume I, p.116; volume II, p. 116.

[49]See volume I, p.107 sqq.

To sum it up, fractions and fractionary numbers seem to have been two different notations expressing the same rational quantity. While fractions were used while working with rationals, fractionary numbers were used to state a rational value[50].

1.1.d Wealth and debt quantities, irrationals, approximate values and so forth

Other types of quantities are manipulated and discussed by our commentator. We will mention them here but leave this area open for further study. In the commentary of verse 30, rules are given to compute with "wealths" (*dhana*) and "debts" (*ṛṇa*) which have been understood as rules of signs[51]. These rules are in a very corrupted dialect of Sanskrit and are difficult to decipher. In the supplement for this commentary we have discussed such quantities as computational entities and not as negative and positive numbers standing alone[52]. Indeed, quantities labeled in such a way belong to the procedure: they do not appear as results.

Approximate (*āsanna*) and exact (*sphuṭa*) measures are considered the quality of approximations discussed. Practical (*vyāvahārika*) computations are opposed to accurate (*sūkṣma*) ones[53]. Bhāskara resorts sometimes to approximate values: this can be done explicitly, and he then specifies the method he uses to take this approximation[54], at other times these values are not explicitly given as approximations and we do not know for sure how they were arrived at[55].

Irrational numbers are discussed and manipulated in several areas of Bhāskara's commentary under the name *karaṇī*. They appear as measures of lengths and areas which cannot be expressed directly and are thus stated through their square values[56]. Some aspects of these handlings of *karaṇīs* are expounded in [Chemla & Keller 2002]. We also intend to probe elsewhere the different understandings of the word and its link with irrational numbers in the Middle East and ancient Greece.

1.1.e Distinguishing values and quantities

It may be helpful, in order to understand Bhāskara's conception of numbers, to establish a difference between the value of a number (*sāṅkhyā*) and the quantity (*rāśi*) it represents. Our hypothesis is that Bhāskara considered that quantities were essentially integers. A number could sometimes be manipulated under such conditions that the expression of its value as an integer was impossible. This would justify and explain how he manipulated rational and irrational numbers. This idea should be a useful guide to Bhāskara's manipulations rather than considered as a definitive statement. I have

---

[50]The case of rational values smaller than 1 being problematic when it occurs, as underlined in the present introduction on p.xxv.

[51]See [Shukla 1976; lxii].

[52]See volume I, p.121 ; volume II, p. 133.

[53]See volume I, p.50.

[54]See volume I, p. 64.

[55]This is for instance systematically the case in BAB.2.11, volume I, p. 57, and discussed in the supplement for this verse commentary, volume II, p. 54.

[56]For instance, in BAB.2.6.cd, volume I, p. 30.

analyzed elsewhere some of the passages that may substantiate this assumption, which remains hypothetical[57].

1.2 Tabular arithmetics?

Only four procedures (squaring, cubing, extracting square and cube roots) mentioned in the treatise are expounded by Bhāskara as depending on the decimal place value notation. One can relate the decimal place value notation to a more general feature of Bhāskara's arithmetics: the use of a tabular disposition for numbers used in a procedure. These tabular dispositions are represented in the "setting down" part of the solved examples of Bhāskara's commentary[58], and sometimes explicitly referred to in words.

1.2.a Classification and Transposition Bhāskara's explanation of a procedure is always grounded on a classification of the entities that it puts into play. To be applied, an algorithm requires a transposition of this classification on a working surface. Thus a Rule of Three has a measure quantity (*pramāṇarāśi*), a fruit quantity (*phalarāśi*), a desire quantity (*icchārāśi*) and a fruit of the desire (*icchāphala*). In arithmetics, this classification is associated with a tabular disposition on a working surface. Thus, in the Rule of Three, Bhāskara prescribes to place the measure quantity on the left, the desire quantity on the right, and the fruit in the middle, all on the same horizontal line. As when numbers are stated with words in a complex way, a silent operation is at work, as a problem is transposed and rewritten on a working surface where it will be used.

In a nutshell, when solving equations[59], inverting procedures[60] or performing a *kuṭṭaka*[61] there is a specific setting, within a table, on a working surface, of the quantities to be used and produced during the procedure.

1.2.b Characterizing tabular dispositions This tabular disposition can be referred to within the procedure itself which can state that one should "move" a quantity (as in rules of proportions involving fractions[62]), or "multiply below and add above" (in the *kuṭṭaka*[63]). Thus a position within a tabular setting is used to indicate, within a given procedure, what operations a quantity will be involved in. This is especially clear in the rules of proportions where "multipliers" are set in

---

[57]See [Keller 2000; volume 1, II.2.].

[58]Summarized in Table 1 on page xviii.

[59]See BAB.2.30, volume I, p. 121; volume II, section V.

[60]See BAB.2.28, volume I, p.118; volume II, p. 128.

[61]The "pulverizer" process which solves an indeterminate analysis problem is one of the classical problems of medieval Sanskrit mathematics. Concerning the process presented by Āryabhaṭa and expounded by Bhāskara see BAB.2.32-33, volume I, 128; volume II, p. 142.

[62]See BAB.2.26-27.ab, volume I, p. 107 explained in volume II, p. 118.

[63]See BAB.2.32-33, references above.

a specific place (on the left in Shukla's printed edition) and "divisors" in another (on the right according to the printed text). Moving quantities can then be an arithmetical operation, as when we invert fractions by moving the numerator to the denominator and vice versa. However, tabular dispositions are local: a disposition changes from procedure to procedure. For instance, in some procedures, a dividend is placed above a divisor (as when fractions are noted) and in others below it (in the *kuṭṭaka*).

1.2.c Various spaces Most complex procedures use several spaces: one where elementary computations will be carried out, one to store a quantity that may be needed later, and a table where quantities arise and are manipulated at the "heart", so to say, of the procedure. However some computations seem to have been performed in no specifically allotted space, or within a place where previous quantities were noted but erased. To sum two given numbers, Bhāskara, at times, states the expression *ekatra*, "in one place". This suggests that the two numbers were erased and replaced by their sum.

1.2.d Computational marks We have seen that a space on a working surface could indicate the operational status of a quantity in a procedure. Marks associated with a given number may have also fulfilled such a role. Thus abbreviations of operations are used in the commentary of verse 28, indicating what operation the number has entered. This allows a mechanical inversion of the operations undergone[64]. In other instances a little round exponent may indicate that a subtraction should be carried out, as in the notation of fractionary numbers[65].

1.2.e Zeros and empty spaces No rule is given to carry out operations with zero in this text, although they can be found in a contemporary treatise authored by Brahmagupta[66]. Could the idea of zero have emerged with the notation of an empty space in the decimal place value notation? Indeed, a *bindu*, a small circle, is used to note empty spaces in the tabular dispositions of quantities in the printed edition of the commentary. In the disposition of a Rule of Five (see examples 11, 12 and 13 in the commentary of verse 26) a *bindu* figures the empty space where the sought result should be placed. Similarly, at the end of example 2 in the commentary of verse 25, $\frac{3}{4}$ is noted with a circle above it, $\overset{0}{\underset{4}{3}}$. This notation also underlines how Bhāskara seemed to avoid stating a result with a

---

[64]See BAB.2.28, volume I, p.118; volume II, p. 128.

[65]Note that the remarks that follow are based on what can be observed in the printed edition of the text. As already brought to light on p. xiv, one needs to be cautious about what such marks testify to. Notwithstanding the editor's own innovations, the existing manuscripts are separated from the text by more than a thousand years.

[66][Dvivedi 1902].

"fraction". Similarly, and most impressively, at the end of the commentary on verse 2, Bhāskara "sets down" the places of digits in the place-value notation of numbers. Each place is noted by a *bindu*.

The name used for zero in Sanskrit, *śūnya*, means "empty". These dispositions thus suggest that the number zero could have evolved from the mark indicating an empty space in the tabular disposition of quantities on the working surface. This however may be an artifact of the edition: the manuscripts we have of Bhāskara's commentary are all in Malayalam script which does not note zeroes with a little circle. Zero is noted as a cross in these manuscripts. In the manuscripts we have consulted no such use of empty spaces can be seen.

1.3   Conclusion

The general impression conveyed by a survey of Bhāskara's arithmetics highlights the importance of the spatial notations of numbers while performing algorithms. The hypothesis of a tabular practice of arithmetics needs, however, to be more thoroughly sustained: it is extremely difficult to distinguish and redistribute what our reflections owe to the innovations and transformations of the modern editions, the manuscripts and finally to Bhāskara's text. Still, this characterization of Bhāskara's arithmetics seems to be worth pursuing, and raises a number of interesting questions: do we have other testimonies of such tabular practices? Are there historical and regional variations of such activities? Is this a specifically Indian way of practicing mathematics, does it bear similarities with traditions of other regions of the world, such as China? Let us hope that such questions will stir sufficient curiosity to impel further probing into them.

Let us now turn to the geometrical aspects of Bhāskara's work.

2   Bhāskara's geometry

We will first examine how Bhāskara defines geometrical figures. In a second section we shall turn to two elements of Bhāskara's geometry: the use of the sine and the existence of false rules to compute volumes. Ideally we would like to recover what was the basic coherence of Bhāskara's geometry, what made possible a continuity from concept to practice. The following is but a local, partial, sketch in this direction.

2.1 Geometrical figures

This section is devoted to a description of the geometrical vocabulary used by Bhāskara. Plane figures are called "fields", *kṣetra*[67].

2.1.a Quadrilaterals Figures with four sides are generically called "quadrilaterals" (*caturaśra*, literally "possessing four edges"; or *caturbhuja*, "possessing four sides").

Figure 1: Quadrilaterals

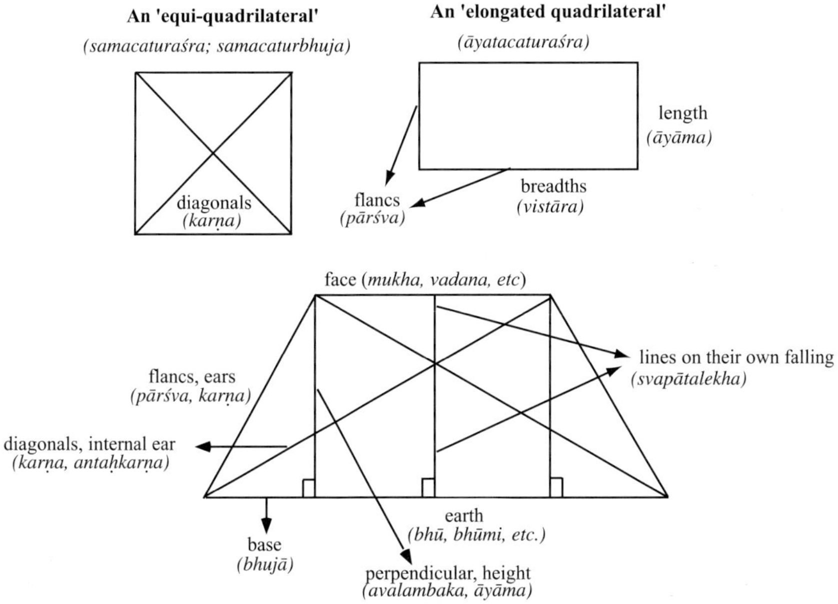

A square is called an "equi-quadrilateral" with equal diagonals (*karṇa*, literally "ear").

A trapezium is defined by the sides that circumscribe it: the earth (*bhū*) parallel to the face (*mukha*), and its lateral sides. They are called "flanks" (*pārśva*) by Āryabhaṭa, "ears" or "diagonals" (*karṇa*) by Bhāskara. It is distinguished from any quadrilateral by the fact that its heights (*āyāma*) are equal. Āryabhaṭa gives a rule to compute the length of the two segments of the height having the point of intersection of the diagonals for extremity. These segments are called the "lines on their own falling" (*svapātalekha*)[68]. The names of the sides of this figure and other quadrilaterals are illustrated in Figure 1.

---

[67] An analysis of the common meanings of the geometrical terms used by Āryabhaṭa and Bhāskara can be found in [Filliozat 1988a; p. 257-258].

[68] All of this is detailed in the supplement for BAB.2.8 in volume II, p. 34.

A rectangle is called an "elongated quadrilateral" (*āyatacaturaśra*). It has a breadth (*vistāra*) and a length (*āyāma*). This pair of terms, when used to refer to sides, may be a way of expressing orthogonality[69]. Bhāskara's interpretation of the first half of verse 9 bestows a central position to the rectangle in geometry. Indeed, he considers that any field can be turned into a rectangle having the same area[70]. Construction wise, this means that one can break up any field and re-adjust the segments into a rectangle having the same area. This strikingly evokes the kind of operations commonly carried out in the *śulbasūtra* geometry. Bhāskara furthermore insists that areas of geometrical fields can be verified with this property[71].

The construction of quadrilaterals is described in the commentary of verse 13. Quadrilaterals are drawn from their diagonals. One can note that inner segments are precisely the elements used by Bhāskara to distinguish and characterize different quadrilaterals.

2.1.b Trilaterals Triangles are called "trilaterals" (*tryaśra*, literally "possessing three edges", *tribhuja*, "possessing three sides"). There are three classes of trilaterals: equilaterals (*sama*), isosceles (*dvisama*), and scalene (*viṣama*). The base (*bhujā*) or earth (*bhū*) is distinguished from the other two sides (*pārśva* or *karṇa*) by the fact that the height (*avalambaka*) falls on it. In the commentary of verse 13 the construction of a trilateral is described using the height and its corresponding base. Again, inner segments appear to be key elements in the characterization and production of a figure. The names of the sides of trilaterals are illustrated in Figure 2 on page xxix.

A right-angled triangle is a specialized field. To name it, its three sides are enumerated: the hypotenuse (*karṇa*), and the two perpendicular sides, the upright side (*koṭi*) and the base (*bhujā*). This is systematically done before applying the "Pythagoras Theorem", given in the first half of verse 17.

---

[69] Please refer to the supplement for BAB.2.9 (volume II, p. 40) for more on this subject.

[70] This is briefly discussed in [Hayashi 1995; p. 73-74], [Sharma & Shukla 1976; p. 43-44] and in the supplement for BAB.2.9, *op.cit.*

[71] The exact reasoning existing behind what Bhāskara calls a verification (*pratyayakaraṇa*) remains elusive, as we will see further on.

Figure 2: Trilaterals

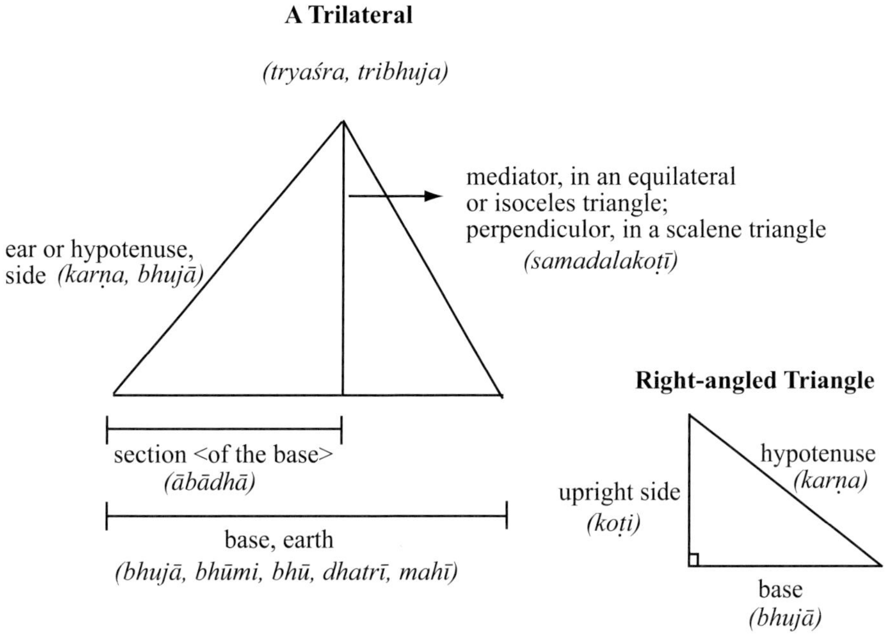

2.1.c Circles   A circle is generally called "an evenly circular ⟨field⟩" (*samavṛtta*) probably to distinguish it from an "elongated circular ⟨field⟩" (*āyatavṛtta*), e.g. an oval. It is defined by its circumference (*samapariṇāhu, samaparidhi*) and its radius, called a "semi-diameter" (*vyāsārdha, viṣkambhārdha*). Bhāskara opposes the circumference of a circle to the disk it circumscribes. This is illustrated in Figure 3 on page xxx.

Uniform subdivisions of the circumference are considered. The most common is the *rāśi*, which is 1/12th of the circumference of a circle. In the commentary of verse 11, Bhāskara considers pair subdivisions of a *rāśi* (1/2, 1/4 or 1/8th of a *rāśi*, that is 1/24th, 1/48th, etc. of the circumference).

In his commentary of the second half of verse 9, Bhāskara introduces the bow field (*dhanuḥkṣetra*) with its arc, arrow and chord[72]. As we will see below, this field is an essential element of Bhāskara's trigonometry. Bhāskara considers also the regular hexagon inscribed in a circle.

---

[72]Se BAB.2.9.cd, volume I, p. 50; volume II, p. 45.

Figure 3: Segments and fields within a circle

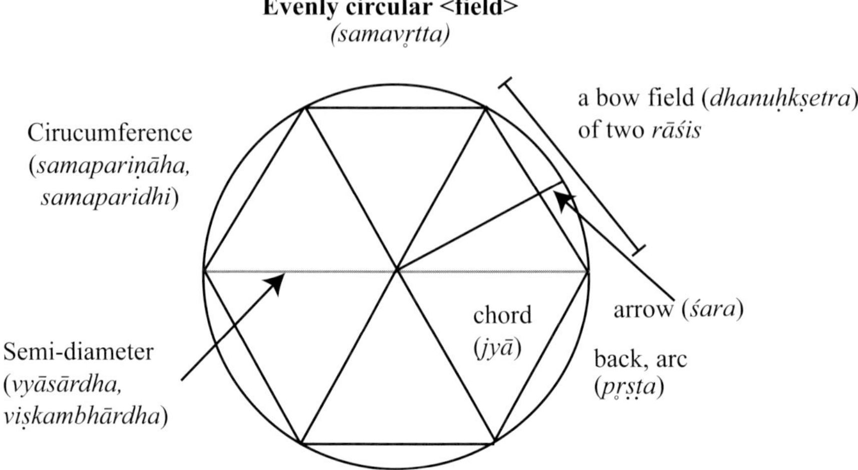

In his commentary of verse 11, Bhāskara considers a figure that can be seen as an ancestor of the "trigonometrical circle". He uses the triangles and rectangles within this field to evaluate half-chords, or sines. This field is illustrated in Figure 4 on page xxxi[73].

In what follows we will have a look at two separate features of Bhāskara's geometry: sine production and the determination of volumes. This commentary provides us with the oldest testimony we know of sine derivation in India. Let us thus briefly have a look at the geometrical context in which they are manipulated.

2.2 Half-chords

In Bhāskara's commentary, we can observe a detailed treatment of the sine. The context in which it is manipulated shows the advantages one had of using half-chords over whole-chords.

A half-chord, (*ardhajyā*) is defined as follows[74]: half of the whole chord (*jyā*) subtending the arc 2α is called the half-chord (*ardhajyā*) of α, see Figure 5 on page xxxii.

The confusion caused by the fact that the expression uses *half the arc of the*

---

[73]The manuscript-diagram reproduced here is a copy of KUOML Co 1712 (47 recto). I would like to thank the library staff and director for providing this copy to me.

[74]This is exposed with more details in the supplements for BAB.2.9.cd (volume II, p. 45), BAB.2.11 (volume II, p.54), BAB.2.12 (volume II, p. 69), BAB.2.17.cd (volume II, p. 101) and BAB.2.18 (volume II, p. 105).

Figure 4: An "ancestor" of the trigonometrical circle, as seen in [Shukla 1976] and in a manuscript.

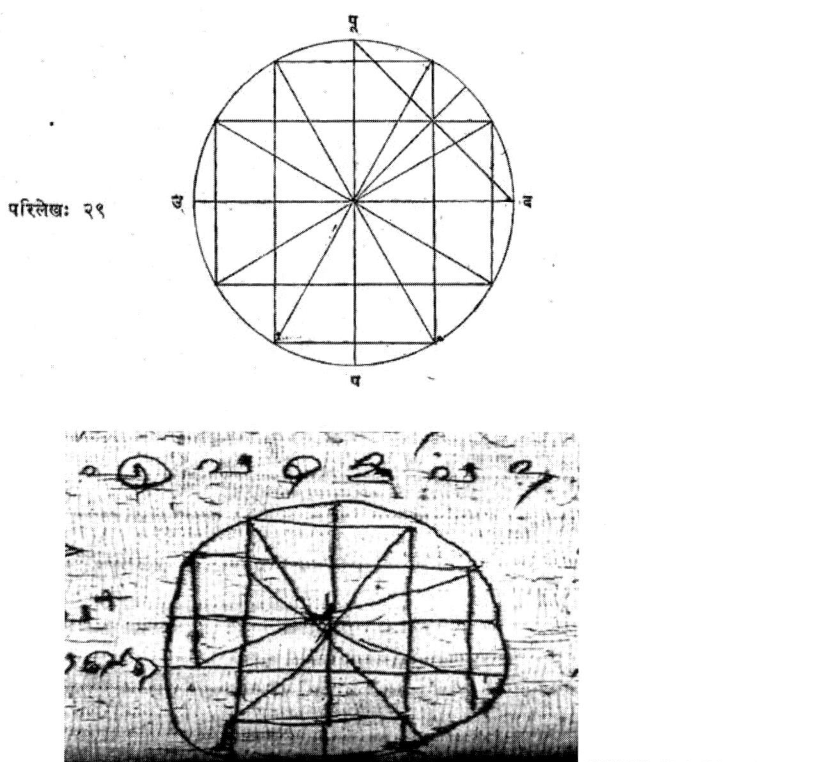

KUOML 1712 Folio 47 recto

original whole chord, when considering the half-chord is furthermore heightened by the fact that Bhāskara often omits the word half (*ardha*) when naming the sine, thus simply calling it *jyā*. This may be the testimony of a transition: the moment when the word slowly loses its original meaning of "chord" to endorse the technical meaning it will have in later literature, that of "sine", "Rsine" specifically. Indeed, the half-chord thus defined is in fact $R \times sin\alpha$, noted here Rsinα. Circles considered by Bhāskara do not have a radius equal to 1.[75]

Bow fields and half-chords are closely intertwined. As in the case of half-chords, the arrow of a 2α bow field is called "the arrow of the arc α" (this can be seen in Figure 5 on page xxxii). This arrow is sometimes called *utkramajyā*. It is defined

---

[75]When discussing $\sqrt{10}$ as an approximate value of π, Bhāskara's staged opponent considers a circle of diameter 1. See BAB.2.10, volume I, p. 50. This stratagem is not used when computing half-chords. The radius commonly in use has the value of the radius of the celestial sphere, 3438 minutes. See BAB.2.11, volume I, p. 57, and the supplement on Indian astronomy, volume II, p. 186.

Figure 5: Arcs, chords, half-chords and arrows

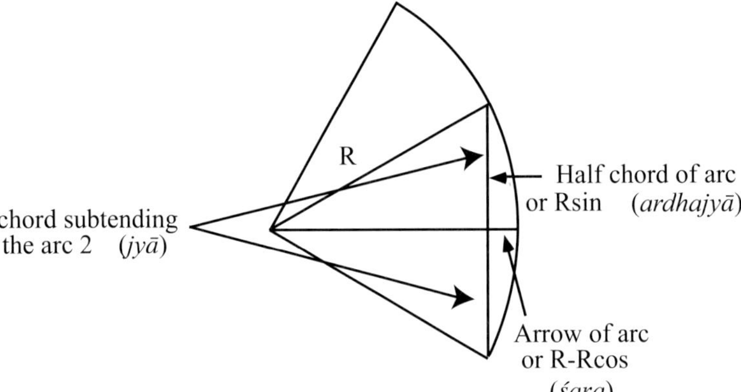

as $R - Rsin(90 - \alpha)$ by Bhāskara. In other words, it is $R - Rcos\alpha$. The cosine is not specifically identified by Bhāskara, however.

In the field worked upon in the commentary of verse 11, which can be seen in Figure 4 on page xxxi, the "Pythagoras Theorem" is used to derive the length of the half-chords, knowing the semi-diameter.

The bow field appears also in several other geometrical contexts which also involve a circle and the "Pythagoras Theorem"[76]. A close look at these problems always brings up a bow-field. This figure thus seems the central locus of trigonometrical calculation in Bhāskara's commentary, in contrast to the right-angled triangle that the English word "trigonometry" recalls.

2.3   Descriptive procedures to compute volumes

If we put aside the piles considered in the verses pertaining to series, three solid figures are described in the *Āryabhaṭīya*: the cube, the sphere and the equilateral tetrahedron. Setting aside the case of the cube, the procedures given in the treatise to compute the two remaining volumes are false[77]. Bhāskara does not seem to realize that these procedures do not give correct results.

As we have mentioned above, the first half of verse 9 is understood by Bhāskara as a statement encompassing all plane figures. I would like to argue here that Bhāskara possesses also an overall idea explaining the nature of solid figures. This idea, additionally, would clarify what sustained Bhāskara's commitment to Āryabhaṭa's mistakes. His vision of solid figures appears through a specific reading he provides

---

[76]Such as BAB.2.17 (volume I, p. 84) and BAB.2.18 (volume I, 92).

[77]For more details see BAB.2.6.cd (volume I, p. 30; volume II, p. 27) and BAB.2.7.cd (volume I, p. 35; volume II, p. 32).

for Āryabhaṭa's rules on solids[78]: he reads them as simultaneously providing an algorithm and a description of the figures considered. These descriptions all tend to view solids as derived from plane figures on which a height is erected. Bhāskara may have believed in a continuous simple link between a plane figure and its associated solid. The reading of the verse giving the rule to compute cubes (Ab.2.3.cd) rests upon the reading of the verse giving the area of the square (Ab.2.3.ab). A cube is constructed from the surface of a square on which a height is raised. The volume of the cube is the product of the area of the square by its height ($V = A \times H$). Similarly, the volume of the sphere is the square root of the area of the circle multiplied by the area ($V = A \times \sqrt{A}$). The volume of the sphere thus seems to be the product of an area with a "height", represented numerically by the square root of the area. The area of an equilateral trilateral is the product of half of the base and a height ($A = 1/2 \, , B \times H$). In continuity with this computation, the volume of the equilateral pyramid is given with the same consideration: half the area multiplied by the height ($V = 1/2 \, , A \times H$). Bhāskara furthermore argues for the "evidence" (*pratyakṣa*) that the volume of the pyramid springs from the plane triangle on which a height is raised.

Thus the continuity between the plane and solid bodies, continuity which would have been insured by the height from which the solid was derived on a plane surface, would be the understanding that Bhāskara had of Āryabhaṭa's false procedures.

2.4 Conclusion

Bhāskara's geometrical practices are manifold. Two activities have been highlighted here. They emphasize the visual character of Bhāskara's geometrical concepts. Indeed, sines are always inserted in a "bow field" volumes are conceived from the area that shapes their base and the height that would be the backbone of the third dimension.

Thus, we have briefly looked at some of the arithmetical and geometrical practices of Bhāskara's commentary. The essential feature being that both *rāśigaṇita* and *kṣetragaṇita* required a working surface on which mathematical objects were noted and worked with. This working surface was represented in the text, and referred to. Bhāskara's "tabular" arithmetics, his use of diagrams in geometry may have not been peculiar to him. Were they included in a larger set of practices belonging to a school, a region, a time? Let us hope that further research will enable us to provide some answers.

We will now turn to the elements of Bhāskara's mathematical activities that shed a light on what he thought of the relations between his arithmetics and his geometry.

---

[78]Below, on p. xl, we will see other reading techniques used by Bhāskara.

3 Arithmetics and geometry

In his introduction of the Chapter on mathematics, Bhāskara introduces a partition
of *gaṇita* in terms of arithmetics and geometry. He states[79]:

> *apara āha- gaṇitam rāśikṣetram*[80] *dviddhā'| (...) gaṇitam dviprakāram
> rāśigaṇitam kṣetragaṇitam| anupātakuṭṭākārādayo gaṇitaviśeṣaḥ rāśi-
> gaṇite 'bhihitāḥ, średdhīcchāyādayaḥ kṣetragaṇite| tad evam rāśyaśritam
> kṣetrāśritam vā aśeṣam gaṇitam|*
>
> Another says: "mathematics is two fold: quantity (*rā'śi*) and field (*kṣetra*)'.
> (...) mathematics is of two kinds: mathematics of fields and mathemat-
> ics of quantities. Proportions, pulverizers, and so on, which are specific
> ⟨subjects⟩ of mathematics (*gaṇitaviśeṣa*), are mentioned in the mathe-
> matics of quantities; series, shadows, and so on, ⟨are mentioned⟩ in the
> mathematics of fields. Therefore, in this way, *mathematics as a whole
> rests upon the mathematics of quantities or the mathematics of fields.*

The particle used in Sanskrit to express "or", *vā*, is non-exclusive. Therefore,
a given procedure can belong to arithmetics, to geometry, or to both. This is
very clearly stated, when below Bhāskara provides a geometrical reading of what
appears to be an arithmetical paradox and then states[81]:

> *evam kṣetragaṇite parihāraḥ/ rāśigaṇite parihārārtham yatnaḥ karaṇīyaḥ/*
> This is a refutation in the mathematics of fields. An attempt should be
> made aiming at a refutation in the mathematics of quantities.

Specifically, in this paragraph, Bhāskara gives a geometrical interpretation of multiplication and division. Geometrically, these operations would allow the transfer
from side to area and vice versa. Elsewhere the multiplication of an area and a
side produces a volume. In other parts of Bhāskara's commentary as well, one can
see geometrical readings of elementary arithmetical operations. Hence, verse 3 of
the second chapter of the *Āryabhaṭīya* gives simultaneously a geometrical and an
arithmetical definition of squares and cubes[82]. Both are expounded by Bhāskara.
In various occasions additions and subtractions are read as the cutting or adding
of segments along a straight line. Dividing by two is taking half of a segment,
and so forth. The *samkramaṇa* operation, a procedure stated in verse 24[83] finds
two numbers, knowing their product and difference. The operation that bears this
name outside of that verse commentary is the part of the procedure that adds or
subtracts a number and takes half of the result. Its geometrical applications[84] rests

---

[79][Shukla 1976; p. 44; lines 15-19] for the Sanskrit; volume I, p. 6; the emphasis is mine.

[80]Reading Shukla's emendation of the text, rather than the *kālakṣetra* of all manuscripts.

[81][Shukla 1976; p. 44; lines 14-15]; volume I, p. 8.

[82]See Ab.2.3, volume I, p. 13-18.

[83]See BAB.2.24, volume I, p. 104; volume II, p. 104.

[84]See for instance BAB.2.17.cd (volume I, p. 84; volume II, p. 101), BAB.2.6.ab (volume I, p.
24, volume II, p. 22).

upon the fact that adding, subtracting, squaring and halving can have geometrical interpretations.

Two operations are of particular importance when examining the relations between arithmetics and geometry: series (*średdhi*) and the Rule of Three (*trairaśika*). We will also briefly look at an ambiguous object in this respect, the *karaṇī*.

3.1 Series

Bhāskara, following Āryabhaṭa, includes series (*średdhi*) in geometry. He nonetheless, gives them an arithmetical interpretation.

Let us note that our two authors define a series by referring to the sequence from which it is derived. For instance, if we consider the sequence of all natural numbers, zero excluded, its first term is 1, its arithmetical reason is 1. The series formed by the progressive sum of the terms of this sequence $(1, 1 + 2, 1 + 2 + 3, \ldots)$ is defined by Bhāskara as "the series having for first term (*mukha*) and reason (*uttara*) 1".

The vocabulary used by Āryabhaṭa in his verses on series is clearly geometrical. The series are described as piles of objects. The *citighana* (the solid which is a pile) is the name of a pyramidal pile, whose tip is made of one object, the second row of three objects, the third of six, etc. Each row of this pile is called *upaciti* or sub-pile. The number of objects per row is one term of the series mentioned above, the series having for terms the progressive sum of natural numbers. The number of objects in the pile is a term of the series of the progressive sums of the objects in each *upaciti*, that is $1 + (1 + 2) + (1 + 2 + 3) + \cdots$ Bhāskara underlines the geometrical character of this series by placing diagrams – and not numbers – in the "setting-down" part of solved examples. Similarly, the series made of the sums of the squares of natural numbers is called by Āryabhaṭa a *vargacitighana*, a solid which is a pile of squares. This solid is represented within a diagram by Bhāskara as a pile of flat square surfaces. The series formed by the successive sums of the cubes of natural numbers is called *ghanacitighana*, a solid which is a pile of cubes. This is represented as a pile of cubic bricks.

Bhāskara, however, in his commentary, substitutes for Āryabhaṭa's words his own vocabulary, which invests the series with an arithmetical reading. The word *upaciti* is glossed as *saṅkalanā* (sum[85]), *citighana* by *saṅkalanāsaṅkalanā* (sum of a sum[86]). The *vargacitighana* is a *vargasaṅkalanā* (a sum of squares). The same occurs in solved problems were the total number of objects (*dravya*) of a pile defined by the number of its rows (*stara*) is sought; in the general commentary, this number is called the value (*dhana*) of the terms (*pada*). It is understood that the terms are those of the series.

---

[85] [Shukla 1976; p. 109, line 18] for the Sanskrit; volume I, p. 100 for the present translation.

[86] [Shukla 1976; p.109 lines 21-22] and volume I; p. 100. This word is also discussed in the Glossary at the end of volume II, p. 197.

Thus, concerning the verses on series, we can see a geometrical situation read in arithmetical terms and reciprocally an arithmetical problem translated geometrically.

3.2 Rule of Three

Even though the commentary of the Rule of Three provides commercial and recreational applications of it (which have an arithmetical flavor)[87], in other parts of Bhāskara's commentary (such as in the commentary of the second half of verse 6, the commentary of verse 8, etc.) it is used to explain geometrical procedures. Bhāskara applies a Rule of Three in geometry to highlight the existence of proportional entities. In this context, it thus seems to provide a mathematical grounding for procedures involving a multiplication followed by a division.

In geometry, indeed, a Rule of Three appears only in relation to what we call similar triangles. The notion of "similar triangles" is not found in Bhāskara's commentary. However each time the properties of such triangles are involved in a problem, Bhāskara quotes a Rule of Three. As a consequence, the Rule of Three may be seen as Bhāskara's way of stating the existence of such triangles[88].

The Rule of Three in this text can also be used to give a new reading of an algorithm. This is clearly the case, for instance, in BAB.2.15[89]. This technique of re-reading a given procedure as a set of known procedures, in India and in China, may have been a method intending to ground or prove the newly read procedure[90].

3.3 Karaṇīs

Karaṇīs are difficult and paradoxical entities in Indian mathematics. They are usually referred to in secondary literature as "surds", but this interpretation remains often problematic when one attempts to analyze what they represent in a given text[91]. I will not demonstrate here how we have arrived at the understandings we propose of these objects in Bhāskara's commentary[92]. Our conclusions only will be presented here, as they will be helpful for the reader.

If one needs the square root of a number N that is not a perfect square, the quantity is called N *karaṇīs*. This expression refers to what we will call "the square root number of N" (N being any positive rational number): $\sqrt{N}$. When such quantities

---

[87]See BAB.2.26-27.ab, volume I; p. 107.

[88]See for instance, BAB.2.6.cd, volume I, p. 30, volume II, p. 27.

[89]See volume I, p. 75.

[90]This hypothesis is brought up also on p. li, below.

[91]See [Hayashi 1977].

[92]A first attempt can be found in my PhD thesis ([Keller 2000; volume 1, II.2.4.5]). This is more precisely outlined in [Chemla & Keller 2002]. We hope to provide someday a full-fledged analysis of the different meanings the word can take and examine its relations with similar paradoxical entities in connection with irrationals in Mesopotamia and in ancient Greece.

are manipulated the value of N is what is used in computations, but $\sqrt{N}$ is in fact the quantity considered. This is the paradox which makes the notion difficult to grasp.

In most of the cases, *karaṇīs* emerge when a computation using the procedure corresponding to the "Pythagoras Theorem" produces a square whose root cannot be extracted[93]. The length of the segment, and not its square, is however needed to solve a problem. Thus quadratic irrationals appear in the computation of the area of trilaterals, the volume of an equilateral pyramid and that of a sphere[94].

As when we compute with square root symbols, when a *karaṇī* number appears, if we want to perform an elementary operation on it using other numbers, we need to transform these other quantities into *karaṇīs*. In other words, we have to put them under the square-root symbol. And to do so, we have to square them. Consequently, under the name *karaṇī* integers (as square roots of perfect squares) and irrationals (as square roots of non-perfect squares) are referred to.

These manipulations of *karaṇīs* may bring us to think that it is an arithmetical entity. In his introduction to the mathematical chapter of the *Āryabhaṭīya*, Bhāskara, however, defines a *karaṇī* operation. He strongly insists that this operation belongs to *kṣetragaṇita* or "geometry". A *karaṇī*-operation, geometrically, is the construction of a square knowing one of its sides. To be more specific, Bhāskara explains that a *karaṇī*-operation is what "makes", in a right triangle, the hypotenuse equal to the other sides. That is, in a right-angled triangle, the square constructed from the hypotenuse has an area equal to the summed areas of the squares constructed from the triangle's two other sides. This may be an etymological pun[95]: the word *karaṇī* is derived from the verbal root *kṛ-* (to make). Thus, when directly associated with the "Pythagoras theorem", a *karaṇī* represents first a geometrical entity, the operation of producing the square of which a given side is known. But it also remains a numerical entity as well, the measure of the side of a square, whose area is known.

In this last case, one can thus understand a *karaṇī* as an indirect way of expressing a measure (that of a length, an area or even a proportion) by using its square. The complexity of this entity perhaps highlights the difficulties Bhāskara himself had in articulating the links between arithmetics and geometry.

3.4 Conclusion: Measuring and numbering

Most of the links between a geometrical problem and its arithmetical reading, as observed in Bhāskara's commentary, arise from a measuring operation. This

---

[93]Bhāskara does not dwell on the links that such quantities bear with the procedure to extract square roots (*vargamūla*, literally: "the root of a square"; for which an algorithm is given in verse 4, see volume I; p. 20.). The "Pythagoras theorem" is given in BAB.2.17.ab, volume I; p. 83.

[94]See BAB.2.6.ab (volume I, p. 24), BAB.2.6.cd (volume I, p. 30), BAB.2.7.cd (volume I, p. 35).

[95]See [Hayashi 1977].

activity is never discussed by Bhāskara. From time to time words derived from the verbal root *mā* (to measure) are used. However measures for the sides of geometrical figures or their areas are given without any explanation. In geometry, when abstract figures such as circles, trapeziums, etc. are considered, no measuring unit is provided. Numbers for the sides are stated in compounds having for last word *saṅkhyā* (number, value) or *pramāṇa* (evaluation, measure). Measuring units appear in "concrete"-like situations where living beings (humans, animals or even flowers) move. They also appear in gnomonic problems. A great variety of measuring units are then put into play. These are listed in a glossary[96].

Series are an exception: a series becomes arithmetized because the objects piled within it are numbered, not measured. But the sum of squares and cubes provides the total area or volume of the objects summed. This is the only case where series and measures are linked.

Hence, arithmetics and geometry are intricate in more than one way. Although Bhāskara may have considered that all procedures could have double readings, one in arithmetics and one in geometry, the unweaving of these links would have often been difficult or even impossible. After the introduction, he never discusses this point again.

In fact, several arithmetical procedures are not bestowed any geometrical interpretation. This concerns the commercial rules (verse 25 and the rules of Five, Seven and Nine[97]), and the rule given in verse 23[98].

Moreover, one wonders why such rules are to be found in a treatise, or its commentary, whose primary subject is astronomy. This puzzling fact raises the question of the relations of mathematics and astronomy in Bhāskara's commentary, to which we turn now.

4   Mathematics and astronomy

At the beginning of his commentary on the *Āryabhaṭīya* itself, Bhāskara explains that the procedures given in the *gaṇitapāda* are stated in order to be applied in astronomy. He specifies that they will be applied in the two following chapters of the treatise, the *kālakrīyāpāda*, or chapter on the measure of time, and the *golapāda*, chapter on the sphere. Bhāskara at this point investigates the links of *gaṇita* in all its generality with astronomy[99]. In some instances, in his commentary on the *gaṇitapāda*, the commentator provides immediately astronomical interpretations of these mathematical procedures. The rules that Bhāskara links to astronomy in the chapter on mathematics are given in Table 3 on page xxxix.

---

[96]volume II, p. 222.

[97]See BAB.2.26-27.ab, volume I, p. 107.

[98]See BAB.2.23, volume I, p. 103.

[99]The text of this discussion can be found in [Keller 2000; volume 1, Annex A] and [Shukla 1976; p. 5-7]. I have also discussed it in [Keller forthcoming].

Table 3: Procedures linked to astronomy in the *gaṇitapāda*

Verse commentary	Procedures
Verse 14	Construction of gnomons, interpretation of the shadow at noon.
Verse 15	Source of light, gnomon and its shadow.
Verse 16	Source of light, two gnomons and their shadows.
Verse 18	Intersection of two circles. Computing the span of an eclipse.
Verse 26-27ab	Rule of Three and the orbit of a planet.
Verse 28	Inversion of a procedure in order to find the time in *ghatis* knowing the Rsine of the altitude.
Verse 31	Meeting time of two planets
Verse 32-33	Pulverizer applied to astronomy in order to produce the longitude of a planet at a given time, or the numbers of days elapsed since the beginning of the *Kaliyuga*.

Two general astronomical topics are discussed in Bhāskara's commentary on the *gaṇitapāda*: The first is concerned with the information that can be deduced from the shadow of a gnomon (longitude, latitude, zenith distance, etc.). The second is linked to the movement of planets (moment of an eclipse, number of days elapsed since the beginning of a given era deduced from the present longitude of a planet, ...).

The mathematical supplements of the concerned verses show that the rules of proportions, "Pythagoras Theorem" and rules concerning arrows, bows and half-chords are applied in astronomical contexts. Similarly the pulverizer seems to have been developed in order to solve the type of astronomical problem that Bhāskara proposes as an illustration. Nonetheless, these procedures are also given in an abstract general mode. The pulverizer is also an "indeterminate analysis" procedure, the span of an eclipse also a problem of intersecting circles.

Obviously, one aim of Bhāskara's commentary in the mathematical chapter is to highlight how rich and diverse the interpretations of a given procedure can be.

The fact that methods can simultaneously be understood as astronomical as well as geometrical may underline how both Āryabhaṭa and Bhāskara considered the celestial sphere as a uniform and quantifiable space. It also justifies a broad understanding of the word *gaṇita*. If it certainly can be translated as "mathematics" inasmuch as it covers a number of specialized technical subjects using such abstract objects as natural numbers and plane figures, it can also mean "computation" or "procedure" which emphasizes the great ranges of uses that these operations can be subjected to. This brief analysis of the links of mathematics and astronomy is but a call for a thorough study of Bhāskara's commentary on the three other chapters of the treatise. How are mathematical procedures applied and referred to in that part of the commentary? Does Bhāskara articulate the link between these two disciplines in a specific way? Was this idea common in the Indian subcontinent in this period, in later times? Let us hope that further research will provide some answers.

In the following we will try to unravel the complex relation linking the commentary and the treatise, highlighting two aspects of the commentarial effort: its crucial role as interpretation of the treatise, and the mathematical work it develops, which cannot be found directly in the treatise itself. Our aim is to show how much we rely on Bhāskara's interpretation to understand Āryabhaṭa.

C The commentary and its treatise

Before we turn to the interaction of Bhāskara's commentary with Āryabhaṭa's treatise, let us reflect on our ignorance of the social context in which such texts were used, and the consequences this has for our approach.

1 Written texts in an oral tradition

India's Sanskrit tradition is usually considered as oral, despite the huge amount of manuscripts and written texts that it has produced. This has to do with the values that this tradition conveys: it has greater respect for orally transmitted (and heard) knowledge (*smṛti*) than for the written one. This commentary is openly a written text. For instance, in the introductory verse to the commentary of the chapter on mathematics, Bhāskara states[100]:

> *vyākhyānam (. . .) adhunā kiñcit mayā* likhyate
> a bit of commentary is now *written* by me

The treatise, on the contrary, is considered as having been transmitted orally by its author. In Bhāskara's words, Āryabhaṭa "tells, says" (*āha*) his verses[101].

Mathematical activity clearly required writing. Thus rules are given to note quantities in order to work with them (see for instance, the rule given in verse 2[102] for the decimal place value notation, or the one quoted by Bhāskara on how to place quantities in order to apply a Rule of Three correctly[103]). Others were offered on how to properly draw diagrams.

---

[100][Shukla 1976; p. 43, line 10] for the Sanskrit edition, volume I, p. 6 for the present translation. Emphasis is ours.

[101]I have chosen to translate this verb by the ambiguous "states", to ease the reading of the text.

[102]BAB.2.2, volume I, p. 10.

[103]See BAB.2.26-27.ab, volume I, p. 107; volume II, p. 118.

However, some other mathematical practices seemed to have been considered as belonging to the oral sphere. Thus explanations and proofs are often referred to, but not given in the written text.

Because we do not know what was the function of written texts, we do not know why Bhāskara's commentary was written (he does not tell us why) and who used it[104]. We also do not know in which way the text itself was read. Indeed, each verse commentary is somewhat autonomous and seems to stand for itself. Reference is sometimes made to another procedure treated in the *Āryabhaṭīya* as if its contents were known and assimilated. Does this mean that one wasn't expected to read Bhāskara's commentary from one end to the other in due order? Was a thematic reading of Bhāskara's commentary prevalent? Clearly, one didn't read Bhāskara's commentary in the way we read books today. Each verse commentary starts by a word to word gloss. Were the verses to be known by heart? If this wasn't the case, then one had to always refer higher up to the verse to understand the movement of the text. Furthermore, the structure of each verse commentary itself does not unfold logically, or in due order, the different steps of a procedure. Some steps may be expounded only in solved examples, others in the general gloss. This also may show that several readings of a same verse commentary were required to understand all of its meanings.

These are but speculations. They highlight our lack of knowledge of the context in which this commentary and the treatise were studied.

Before turning to Bhāskara's work as an interpretation of the *Āryabhaṭīya*, let us look at what he tells us of his role as a commentator.

2 Bhāskara's point of view

2.1 What a good rule is

Some notations give us a brief idea of Bhāskara's point of view on the verses and conversely on what his mission as a commentator is. According to him, the rule given in the treatise is a seed (*bīja*) that the commentator expounds[105]. We also know that verses should state general characterizations, illustrations being the realm of a commentary[106]. He also explains that rules can be given unsuspected purposes and meanings[107]; rules may be difficult to understand, but that is be-

---

[104]K. V. Sarma's study of the Kerala school of astronomy ([Sarma 1972]) suggests that texts were copied in order to travel and instruct astronomers from everywhere in the subcontinent. In the case of Kerala, this was the privilege of a caste of princely astronomers. However, we do not know if this story is specific to Kerala or can be extended to the whole of the Indian subcontinent.

[105]See BAB.2.26-27.ab, volume I, p. 108, and for the same idea BAB.2.11, volume I, p. 58.

[106]See BAB.2.2, volume I, p. 12.

[107]See BAB.2.9.ab, volume I, p. 44.

cause they require an interpretation[108]. Bhāskara often quotes the *Mahabhāṣya*[109]. Although the structure of his commentary has nothing to do with Patañjali's, this work seems to have the status of a model.

Thus the principal objective of Bhāskara's commentary is to give an interpretation of Āryabhaṭa's verses.

2.2   Defending the treatise

Bhāskara's commentary is, in the general part of each verse gloss, usually structured as a dialog, where objections are raised and replied to. Most of the staged objections seemingly argue in favor of Bhāskara's reading and interpretation of the verses. Objections often propose alternative readings that are discussed and rejected. Sometimes Bhāskara does not stage a dialog but just assesses alternative readings and discusses them. The commentary of verse 10 is thus a long refutation of the approximation $\sqrt{10}$ for π. Thus, sometimes these dialogues represent a debate on the validity of the rule. Commonly however, they highlight that the verse contains all the essential elements of the procedure it describes. In that case, the explanation of the first step of an algorithm is followed by a staged question whose answer can be found in the verse and explains the second step of the algorithm and so forth. If there is a division, what is the dividend and what is the divisor? etc. The "general commentary" to verse 4 can be referred to as a good example of coherent dialogue[110]. Sometimes the answer is not in the verse, in which case Bhāskara needs to supply an operation or a condition.

The commentator has to show that the verse is consistent not only syntactically but also in respect to the mathematical operation it describes. Expounding Āryabhaṭa's verse will thus bring Bhāskara to weave his own interpretation and mathematical skills into our understanding of it.

This is Bhāskara's idea of his commentary. How does he, practically, put it into action?

3   Bhāskara's interpretation of Āryabhaṭa's verses

Let us first look at how Bhāskara provides a non-ambiguous meaning for Āryabhaṭa's verses. We will then look at how he transforms Āryabhaṭa's rules into mathematical procedures.

---

[108]See BAB.2.10, volume I, p. 51.

[109]An infamous grammatical commentary authored by Patañjali, that all educated Indian Pandit knew. See [Filliozat 1988b].

[110]BAB.2.4 in volume I, p. 20 sqq.

3.1 Standard techniques of sūtra interpretation

The history of mathematical commentaries in Sanskrit still needs to be written[111]. Nonetheless, the inclusion of numerous linguistic discussions and quotations seems to be a striking specificity of Bhāskara's commentary.

Let us see how, while doing so, he generates his own specific reading of Āryabhaṭa's verses.

3.1.a Grammatical analysis removing ambiguities Sanskrit is a language characterized by a great polysemy. It is also a language that uses declensions. Inside a compound the syntaxal link of the words will not be indicated. Such features will thus be used to, as Umberto Eco expresses it, "open the door" of the many different readings a sentence can be subjected to.

The commentary proceeds in a systematical fashion: it glosses each word of the verse. Often the expounded expression does not present any problem. The whole of the verse however is always thoroughly analyzed, ambiguous and non-ambiguous phrases are all considered on the same level. Most of the syntactical ambiguities of a verse arise when compounds are employed. For instance, in verse 32 the compound, *adhikāgra* is analyzed by Bhāskara in two different ways. This produces two distinct understandings of the compound: with one analysis it is understood as meaning "the greater remainder"; with the second analysis it means "a big number". Indeed, not only is the word *agra* given two distinct meanings (once as "remainder", once as "number") but its syntactical link to the word *adhika*, "first" is also modified. In the first case, both words are taken as adjectives, and the whole compound is considered as a nominalized adjective (*bahuvrīhi*). In the second case they are perceived as the apposition of two nouns (*karmadhāraya*). In other cases the syntactical link symbolized by a case (genitive, instrumental...) may be discussed.

This double reading in turn modifies the procedure to be applied. In the case of verse 32-33, two different algorithms are read into the verses[112]. Compound analysis is one of the striking moments where the commentator inputs his own interpretation of the verse he is glossing.

3.1.b Substitution of words and lists of synonyms Bhāskara, occasionally, exchanges the words used by Āryabhaṭa with his own terminology. When a whole phrase is expressed in a new fashion, this is announced within the commentary by the expression *iti artaḥ* (the meaning is) or *iti yavāt* (to sum it up). Bhāskara also provides lists of synonyms for the technical terms that Āryabhaṭa employs. The

---

[111]We have noted above (Introduction, p. xvi), the lack of reflection on the commentarial genres in India. Consequently, few studies have been conducted on the question of mathematical commentaries in Sanskrit: [Jain 1995] and [Srinivas 1990] represent first attempts in this direction.

[112]See BAB.2.32-33, volume I, p. 128; volume II, p. 142.

function of word substitution is often to remove the ambiguities that puns and the like would have introduced in the verse. For instance, when commenting upon verse 4, the word *varga*, "square", used to refer to a place in the decimal place value notation, is glossed by the word *viṣama*, "uneven". We have noted above how the reference, within the verse, to a "square place" added a mnemonic flavor to it, but rendered the signification of the verse ambiguous since both square numbers and square places were named by the same word. By changing the typology, Bhāskara intends to unwind the ambiguity and clarify the steps of the procedure at stake. In some instances, the substitution of a word by another aims at giving a new interpretation of the verse altogether, or at least at broadening its meaning. For instance, the first half of verse 6 deals with the areas of trilaterals (*tribhuja*). Literally, Āryabhaṭa's verse seems to be concerned with equilateral and isosceles triangles: a mediator (*samadalakoṭi*) is multiplied by half the base (*bhuja*). Bhāskara by glossing the word used for "mediator" with one meaning "height" (*avalambaka*) confers a greater generality on the verse. Sometimes word substitutions do not seem to be significative. For instance in the commentary of verse 6 the word used for trilaterals, *tribhuja* is glossed by a synonym *tryaśra*. This may be to avoid the confusion of *bhuja* (side) with *bhujā* (base). But it could also be a way to underline the diversity of the existing vocabulary, its variations from region to region or the disparity between the vocabulary in use at the time of Āryabhaṭa and what was coloquial for Bhāskara.

One can thus see how, by helping one to read and understand a verse word for word the commentator, also, gives his own interpretation of it. In turn, by giving his own understanding Bhāskara sheds light on some of the mathematical functions of the ambiguities and puns within the verse, giving us a glimpse off what could have been a typical way of composing technical verses.

3.1.c Elaborating a technical vocabulary As he comments on the words used by Āryabhaṭa, Bhāskara underlines and insists that the vocabulary employed is technical. He himself often specifies and elaborates the vocabulary. For instance, Bhāskara discusses whether a literal, etymological meaning should be attributed to a word rather then a conventional one. Hence both the word used for the geometrical square, *samacaturaśra* (lit. equi-quadrilateral) or the one for the height of a triangle *samadalakoṭi* (lit. an equally halving height) should not, according to our commentator, be understood with their literal meanings[113]. In order to assert such arguments, he points out the restrictions that a literal meaning would bestow on the rules. Thus he explains that some "equi-quadrilaterals" are not squares, or that some triangles do not have "equally halving heights". So that as he justifies his technical understandings of Āryabhaṭa's words he also deepens his characterizations of the objects at hand. This involves elaborating and specifying the existing vocabulary. Thus Bhāskara suggests to label a geometrical square a "*samakarṇasamacatura ´sra*' (equi-diagonal-equi-quadrilateral). Or, the names of

---

[113]See BAB.2.3.ab, volume I, p. 13 and BAB.2.6.ab, volume I, p. 24.

three categories of triangles are given: *sama, dvisama* and *viṣama* for, respectively, equilateral, isosceles and scalene trilaterals. As he elaborates and justifies the use of a technical vocabulary, Bhāskara in fact *defines* the objects such words refer to. In several cases, the elaboration of definitions is interspersed with quotations from grammatical works that consolidate and confirm the need to use a technical vocabulary.

Hence, by justifying Āryabhaṭa's choice of words, Bhāskara brings to light how several understandings could arise from the verses. And in the same movement, he also highlights his own choices. As we see him elaborating mathematically the signification of these words, we can actually observe him inserting his own mathematical input into the text.

3.1.d Supplying, connecting, repeating and omitting Several diverse and specific reading techniques are put forward by Bhāskara. This can involve adding words or omitting some, connecting words in unsuspected matters and bringing in quotations from other verses of the treatise. Let us have a brief look at these methods.

One aspect of the word to word gloss of Āryabhaṭa's verses has to do with filling the gaps, so to say, of the elliptic formulation they contain[114]. This is sometimes done literally, as Bhāskara supplies "the remaining ⟨portion of⟩ the sentence" (*iti vākyaśeṣa*). For example, while commenting on verse 5, Bhāskara is the one that tells what the rule produces, a cube-root (*ghanamūla*)[115].

Furthermore, in an elliptic verse, a same word can have two syntacically functions. For instance the word *ghana* used in the second half of verse 3 can be read both as being the subject of the verse, meaning then the "volume" (of a cube), but also as the qualifying adjective of "what has twelve edges", and thus mean "solid". Bhāskara indicates this by supplying it twice, thus redistributing it to each word[116]. This doubling of the function of one word is possible because the traditional order of words is modified in an elliptical sentence[117].

In some cases the commentator connects (*sambandh-*) the words of a verse because the natural order of the sentence has been lost. Similarly, the action of calling a word used in a former verse (*anuvṛt*), a traditional commentarial technique in Sanskrit literature is sometimes applied. Thus, when commenting on the second half of verse 3, Bhāskara recalls the area of a square, evoked in the first half of the verse. This is the basis on which he constructs the computation of its volume, and probably also the cube itself[118].

---

[114]The concise aspect of Āryabhaṭa's verse is briefly described above on p. xvii.

[115]See BAB.2.5, volume I, p. 22.

[116]See BAB.2.3.cd, volume I, p. 18.

[117]Sanskrit being a language with declensions, word order does not have much importance. Nevertheless, there is a common place for words in a sentence, which the elliptic form disturbs.

[118]See BAB.2.3.cd *op.cit.*

Alternatively, the commentator wanting to add specifications that are not contained in the verse can gloss expletives. Indeed, meter requirements often introduce words that have almost no meaning. If the commentator can briefly state that function by employing the expression *iti pādapuraṇe* (⟨is used⟩ when filling the verse), we sometimes understand such words as introducing an inflexion or a prescription. Thus the regularity of the cube is inferred from the word *tathā*, which means "then"[119]. To further justify some of his interpretations, Bhāskara sometimes refers to tradition (*āgama*). This is paradoxical, since we can elsewhere catch him setting aside an argument precisely because it relies on tradition and not on established (*sādh-*) facts. All of this informs us that if supplying words and meanings to a verse was a natural phenomena in a commentary, such insertions still needed to be justified.

Finally, rather than supplying words, the commentary sometimes suppresses them from a verse in order to convey a new meaning. This is the striking technique used in Bhāskara's commentary of verse 19. While giving an interpretation of the three first quarters of this verse, Bhāskara in his own words "connects, unites" (*saṃbandha, pratibaddha*) some words while omitting others. This kind of "reading by omission" is possible because of the way the verse is composed. It is made of an ordered list of operations, each operation given by one word. If one reads the whole list furnished in the first three quarters of verse 19, no correct rule comes out of it. It is only by skipping one word/operation that appropriate rules can be read into it. This technique enables Bhāskara to read in this portion of the verse no less than four different rules on series[120].

Thus Bhāskara uses specific reading techniques when giving an interpretation of Āryabhaṭa's verses. This uncovers the existence of specialized reading methods which most probably respond to equally specialized ways of elaborating *sūtras*. Additionally, it highlights our indebtedness to the commentator. For indeed, we do not know these different reading tools, nor do we know how to detect verses that would require specialized readings. *Sūtras*, probably, were composed in order to be open to such multiple readings and interpretations. Thus, Bhāskara's commentary discusses alternative interpretations of a same verse. A diachronic study of Āryabhaṭa's commentators would probably bring a greater diversity to light.

3.2   Conclusion: Glossing is giving an interpretation

We have seen how at every step of what appears to be a linguistic, word by word gloss of the verse, Bhāskara can, and indeed does, provide his own interpretation of a rule. He uses a surprisingly diverse number of tools to do so. This art of interpretation culminates in his interpretation of the two last verses of the mathematical chapter. Bhāskara provides, in this case, two separate interpretations of

---

[119]See BAB.2.3.cd, *op.cit.*
[120]For more details see BAB.2.19 in volume I, p. 93; volume II, p. 107.

verses 32-33. To do so he uses almost all of the different techniques that we have specified above.

Bhāskara, to expound the seed in Āryabhaṭa's verse, used special reading methods. This raises a certain number of questions, which provide as many programs of future research. Firstly, up to what point was Bhāskara's text integrated into the larger tradition of commentarial literature in Sanskrit? Indeed, many of the reading techniques employed seemingly belong to this tradition. And Bhāskara appears as an author who can quote from some of this literature. Additionally, these technical readings also highlight the technical aspect of the composition of the verses themselves, the fact that they were probably written to be read in such a way.

Let us now turn to the specific mathematical aspect of his inputs in the commentary.

4 Bhāskara's own mathematical work

A certain number of specifications are systematically needed for the elusive rules provided by Āryabhaṭa. Even when the meaning of a verse is clear, it just gives the core of a mathematical process, not the details. Thus, when unraveling all that is necessary to apply the rule, Bhāskara's mathematical input once again becomes important. Let us examine here how he transforms the instructions given in a verse into a procedure.

4.1 Identifying mathematical contexts

Part of the elliptic character of Āryabhaṭa's verses springs from the fact that the mathematical context in which a procedure is applied is not given in the verse. By mathematical context we mean the type of mathematical objects the procedure is concerned with, the question it gives an answer to, what is required to apply it, the mathematical subjects in which it can be applied, etc. Bhāskara's commentary will, systematically, have to clarify all of these aspects as he transforms the precepts of the verse into an algorithm.

By indicating what are the requirements needed to apply a rule, Bhāskara often completes and extends Āryabhaṭa's *sūtras*. For instance, the second half of verse 6 gives a rule to compute the volume of an equilateral tetrahedron: multiply half the area of the triangular base of the pyramid by its height[121]. In his general commentary, Bhāskara extends the rule and the procedure by showing that once one knows the length of a side of the tetrahedron, one can compute its volume. To do so, he shows how the height of the tetrahedron is derived from one of its sides. Bhāskara here wants to specify the context in which the rule can be applied. To

---

[121] For the rule and commentary, see BAB.2.6.cd, volume 1, p. 30. This rule is incorrect, and discussed in volume II, p. 27. This fact is also briefly discussed above on p.xxxii.

do so, he should list what are the elements required to apply the procedure. But in this case he does more than that. He actually works out mathematically what is the simplest context required. Indeed, his extension reduces what is required to be known before applying the rule.

Sometimes the subject a procedure deals with is not immediately apparent from the rule itself. For instance, to understand the rule given in verse 16, one needs to know in what case it applies. It is Bhāskara that describes this context, where two gnomons have shadows cast by the same source of light[122].

Most frequently, however, it is in the list of solved examples that the context is described implicitly. Let us take the applications of the Rule of Three. The list of solved examples that follows the commentary of Ab.2.26-27.ab which states the rule, starts by abstract arithmetical problems[123]. The first example uses integers, the second fractional numbers. The following examples show how the rule can be applied in "fun" situations (that of a snake gliding into its hole for instance). To this should be added the applications of the Rule of Three in geometrical and astronomical problems that can be found elsewhere in the commentary. Thus Bhāskara provides a wide range of contexts for this rule. Solved examples give us a standard problem to which the rule gives an answer: the versified formulation highlights what should be known beforehand, the mathematical situation in which a procedure makes sense and what the algorithm produces, all together.

4.2   Ordering the steps

According to Bhāskara, Āryabhaṭa's rules only express the seed of an algorithm. This metaphor can be understood as referring to the fact that often the instruction in the verse gives the core operations of a procedure: it will not give all of them or be specific about their order. This is especially clear with iterative processes. For example, verses 4 and 5 state the beginning of the procedure in their last quarters[124]. Indeed, the algorithms they describe are to be repeated an indefinite number of times. The rule insists on this fact by first stating the last operation to be performed. The commentator is left with the task to put back onto its feet a procedure that was stated upside down.

Often one needs to refer to the *karaṇa* ("procedure") part of solved examples, to understand exactly what are the different steps of an algorithm. Solved examples underline the alterations a procedure undergoes according to the objects that are input (be it different kinds of quantities or geometrical objects). For instance, the solved examples of the commentary of the first half of verse 6 explain how to compute the area of a triangle knowing its sides in the case of equilateral, isosceles and finally scalene triangles. In the last case, the procedure to find the

---

[122]See BAB.2.16, volume I, p. 79; volume II, p. 92.
[123]See BAB.2.26-27.ab, p. 109 sqq.
[124]See BAB.2.4-5, volume I, p. 20-22 ; volume II, p. 15.

height of the triangle is different and slightly more complicated than in the first two cases[125]. The progressivity of problems is also striking in the list of solved examples concerning the Rule of Three[126]. As mentioned above, the first example applies a Rule of Three with integers. The following one applies it to fractional numbers. This implies changes in the steps of the procedure itself. The following examples show how the rule can be applied in "fun" situations. In this case, one needs to analyze the problem to "find" where the Rule of Three is hidden and should be applied.

4.3 Standard questions

The versified problems of the list of solved examples set a standard context for a procedure to be applied: the mathematical subject concerned, the prerequisite items and what is sought, all is summed up in the way such problems are formulated. In the case of three specific procedures, this fact seems to be generalized: their application in all contexts seems to be subject to a standard reformulation of the problem concerned. This means that there exists a coded question or phrase that, when used, brings the procedure into the light, indicating that it should be applied, and specifying what are the elements that will be employed in the process. One of these three procedures we will not describe here because it has a very local application: this is the way one computes the "witty" quantity in the pulverizer process[127]. Let us look at the case of the "Pythagoras Theorem" and the Rule of Three.

4.3.a "Pythagoras Theorem" The right-angled triangle in Bhāskara's text, as in most medieval Sanskrit literature, is singled out and named by giving the list of the names of its three sides: the base (*bhujā*), the upright side (*koṭi*) and the hypotenuse (*karṇa*)[128].

In the first half of verse 17, the "Pythagoras Theorem" is given within a right-angled triangle[129]. Consequently, when Bhāskara wants to apply a "Pythagoras Theorem", he will identify the right-angled triangle involved by renaming the segments of the figure it is inserted in. This is especially clear, because systematically and repeatedly done, in the procedure described in BAB.2.11. In this verse commentary, Bhāskara considers "bow fields" within a circle. From being chords, arrows and semi-diameters these segments become upright sides, bases and hypotenuses as he tries to compute their lengths[130]. This naming highlights a

---

[125]See BAB.2.6.ab in volume I, p. 24.

[126]See BAB.2.26-27.ab, *op. cit.* and [Keller 1995] on this subject.

[127]The reader can refer to [Keller 2000; volume 1, I.5.5.c] and to volume I, BAB.2.32-33, p. 128 and its supplement in volume II, p. 142.

[128]This has been described above, on p. xxviii.

[129]See BAB.2.17.ab in volume I, p. 83; volume II, p. 100.

[130]See BAB.2.11 in volume I, p. 57; volume II, p. 54.

property of the figure (it contains right-triangles) and, simultaneously, states the
way the procedure will be applied.

4.3.b The Rule of Three   The Rule of Three has two characteristics. One we
have already observed: it uses a categorization of quantities[131].

Indeed, a Rule of Three has a measure quantity (*pramāṇarāśi*, M) that produces a
fruit quantity (*phalarāśi*, F) and a desire quantity (*icchārāśi*, D) for which we want
to know the fruit of the desire (*icchāphala*). The fruit of the desire is obtained by
multiplying the desire by the fruit and dividing by the measure ($\frac{D \times F}{M}$). Secondly
it is formulated by a standard question. This formulation of the Rule of Three can
be translated as follows:

> If with M, F is obtained. Then with D, what is obtained? The fruit of
> the desire is obtained.

In the first conditional clause the elements of the ratio are known, in the second
one, the unknown element is introduced. This formulation, which requires a proper
identification of the quantities, indicates the operations to be followed (we know
what should be multiplied and what should be divided) and at the same time
the mathematical property that links them (if the ratios were not equal we would
not introduce the rule). Thus, as in the case of the "Pythagoras theorem", a
Rule of Three stresses a mathematical property and introduces a computation
simultaneously.

In his commentary of verse 14, Bhāskara introduces this standard formulation
(*vācoyukti*) that both identifies and states a Rule of Three[132]. When such formulations arise in the translation we have always indicated it in footnotes. However,
if it is the most common way to apply a Rule of Three, the standard question
does not always need to be spelled out. When four quantities are involved, each
quantity only needs to be identified with one of the categories that the Rule of
Three puts into play. As when applying the "Pythagoras Theorem", items will
be given new names. For example, verse 10 gives for a known diameter the approximate circumference of a circle[133]. Bhāskara explains that with this verse, the
diameter or the circumference of any other circle can be found. He writes that
either the circumference or the diameter can be "desire quantities". Even though
he doesn't add anything else, this information is sufficient: one can deduce that
the circumference and diameter given in verse 10 will then be the "measure" and
"fruit" quantities in a Rule of Three. Thus by the use of one word, a Rule of Three
is brought to light.

---

[131]See above on p. xxiv. One can also refer to [Keller 1995], and to BAB.2.26-27.ab in volume
I, p. 107; volume II, p. 118.

[132]See BAB.2.14 in volume I, p. 70, and for an explanation of the Rule of Three applied in that
portion, volume II, p. 78.

[133]See BAB.2.10, volume I, p. 56; volume II, p. 47.

The standard formulation of a Rule of Three stands as a specific syntactical leit-motiv. Even without renaming quantities, when they are woven into the standard formulation one knows what are the ratios at stake and what computation should be carried out.

4.3.c Conclusion A textual practice that is simultaneously a mathematical action has been highlighted: the attribution of technical names, be it the sides of a right-angle triangle or the typology of the quantities in a Rule of Three. Pulling one of these names into a context which was described without them was probably Bhāskara's way to underline a mathematical property. The standard formulation of a Rule of Three, has the same function. These two methods are syntactical in nature. Because they perform two functions simultaneously they can be seen as technical and concise devices.

4.4 Explaining, proving and verifying

Different understandings of Āryabhaṭa's verses are often presented and corrected by Bhāskara within the fictitious staged dialogues of the commentary. Indeed, the commentator's intention is to justify his understanding of the treatise. To carry this out, he can juxtapose different types of argumentations as if he was trying, by all means possible, to convince his reader of the correction of his interpretations. The justification of a verse, we have seen, is not always strictly mathematical. Thus linguistic discussions of words or the syntax of the verse can be at stake rather than its mathematical contents, although it is often difficult to separate one from the other. Sometimes also, even though the mathematical matter is at play, it is not discussed but justified by a meta-mathematical justification: the reference to the authority of tradition (*āgama*). Hence, the quotation of unreferenced verses that seem to spring from an oral tradition, seemingly have a sufficient authoritative value. For instance, in the commentary of the first half of verse 6, the fact that the height in an isosceles or equilateral triangle is also a mediator is presented through a traditional verse. The validity of the statement is not discussed, the simple authority of ancestral knowledge serving as grounding for its exactitude[134]. However, Bhāskara himself can sometimes discard tradition and require a justification instead. Thus he writes in his commentary of verse 10, answering to objections[135]:

> *atrāpi evāgamaḥ naivopapatiḥ . . . cetad api sādhyam eva*
>
> In this case also, it is merely a tradition (*āgama*) and not a proof (*upapatti*) (. . .) But that also should be established (*sādhya*).

And indeed, Bhāskara does feel compelled at times to explain, or even prove, the mathematical procedures he is discussing. Bhāskara thus uses a vocabulary

---

[134]See BAB.2.6.ab, volume I, p. 24.
[135][Shukla 1976; p. 72; lines 15-19] for the Sanskrit; volume I, p. 52.

for what may be explanations, proofs or demonstrations[136]. Thus words derived from the root *pradṛś-*, as *pradarśana* are used in the commentary. Their basic signification, as in English, is "pinpointing", "seeing", but they can also be used (as in the commentary of the second half of verse 17) in the sense of 'explanation'. Bhāskara also employs the words derived from the verbal root *pratipad-*, such as *pratipādana*, which also means "explanation". This is for instance used in the commentary of verse 8 when a mathematical grounding for the rule is evoked. Finally the word *upapatti*, which in later commentaries will be used systematically for what we can name provisionally "the Indian kind of proof", is used several times in the text. In another range of meanings, Bhāskara uses two terms that are linked with attempts to persuade and justify. In his commentary on verse 9 Bhāskara evokes a *pratyāyakaraṇa*. This word is commonly translated as "verification" although its literal meaning is "producing conviction". Finally, part of Bhāskara's works involves putting forward refutations, *parihāra*.

What are the reasonings that Bhāskara named in such a way? Let us note, that we have very few testimonies of them in the text itself. Many of the explanations seem to have been oral. Diagrams would be the only trace left of them in the written text. In certain cases, we have traces of a reasoning but they are fragmentary or remain puzzling. Some of these reasonings are part of the "general commentary"; others are put forth or intertwined with solved examples. Perplexing as these reasonings can seem, this is our hypothesis: An explanation or proof consisted in rereading a procedure by an independent algorithm, that would furthermore give a mathematical grounding to it[137]. For instance, the computation described in verse 15 describes a multiplication followed by a division. We think that when Bhāskara gives a new reading of this rule, with the help of a Rule of Three, his intent is to ground it mathematically, to justify it. If we hope to sustain this argument elsewhere, we have felt that to state this hypothesis here might help the reader in understanding some of the reasonings that can be found in the commentary. This hypothesis does not solve our uneasiness, as we read Bhāskara's verifications[138], or his long "refutation" of the use of $\sqrt{10}$ as an approximation of π[139].

To sum up, if Bhāskara's way of demonstrating or explaining is in no way Euclidean – and therefore remains enigmatic – on the other hand his effort to convince or persuade his reader cannot be denied.

---

[136]Let us not fix the terminology yet, as this is but the premise of a reflexion on the modes of justifications in Indian mathematics. See BAB.2.15 (volume I, p. 75; volume II, p. 89), BAB.2.14 (volume I, p. 70; volume II, p. 78), BAB.2.17.cd (volume I, p. 84; volume II, p. 101), BAB.2.8 (volume I, p. 37; volume II, p. 34).

[137]A first attempt at describing and qualifying these reasonings can be found in [Keller 2000; volume 1. 1.8.3; 8.4; 8.7]. We intend to discuss this in a forthcoming article.

[138]See BAB.2.9.ab, volume I, p. 42, and BAB.2.25, volume I, p. 105. One can refer to [Keller 2000, volume 1, 1.8.4] for an attempt at analyzing the difficulties of such reasonings.

[139]See BAB.2.10, volume I, p. 50. Once again [Keller 2000. volume 1, 1.8.8] attempts a description of the reasoning set forth here.

4.5 Conclusion: solved examples are more than just illustrations

Among the mathematical work that characterizes the practice of the commentator, the solved examples are certainly the most important. They are not mere illustrations of the procedures. They highlight standard formulations of the problems a procedure can solve, explain the different variations that a procedure can undergo and sometimes serve as counterexamples and illustrations of the limits of the rule itself. Furthermore, the "setting down" part of the solved examples is especially precious as it gives us an insight into the concrete way mathematics was practiced.

What was a mathematical commentary in VIIth century India?

This Introduction has thus but briefly described all the different ways in which Bhāskara has expounded the mathematical seed provided by each one of Āryabhaṭa's rules. If the emphasis has been on the method rather than on the contents, the variety of subjects and objects that treatise and commentary put into play can be discovered by reading the translation.

Bhāskara's text provides a first testimony on how one practiced mathematics in VIIth century India: diagrams were drawn, numbers noted in tabular forms and then manipulated.

The text handed down to us furthermore presents a specific conception of what a technical mathematical text is. We have thus seen that there existed specialized methods for a commentator to read the mathematical verses of a treatise. These readings required in fact a continuous flow of mathematical, grammatical and semantic input from the commentator. Commentarial activity, as Bhāskara's text testifies of, ranged from giving a mathematical grounding to a rule and providing a list of solved examples to it, to explaining an uncommon use of a word or explicit syntactical links in an odd compound. How peculiar was Bhāskara's commentary in the Indian context? Was it a typical mathematical commentary? Was it something totally new? Did it influence later mathematical commentaries? In what way did it belong to the larger genre of VIIth century Sanskrit commentaries? I hope that further research will unfold some answers to these questions.

On the Translation

This translation aims at giving access to a mathematical commentary in Sanskrit, to those unfamiliar with this language. I have tried to make as literal a translation as possible in order to point out the way reasonings have been expressed. Some locutions may not correspond to those used in English. However, I have, when I did not think it was relevant to the mode of thinking, translated Sanskrit phrases into what seemed more colloquial English. Most of Bhāskara's verse commentaries come with mathematical supplements which aim at helping the reader understand the technical discussions and computations they contain. These are included in the second volume of the book. Finally, the translation of Āryabhaṭa's verses is given according to Bhāskara's understanding of them. Occasionally, this specificity is noted in footnotes.

1   Edition

This translation was made using K.S Shukla's edition of the *Āryabhaṭīyabhaṣya* published in 1976 by the INSA in New Delhi. The pages and line numbers, which are given in the right margin of the translation, refer to this edition. Geometrical figures have been counted as one line.

Sometimes the editor, K.S. Shukla, has substituted manuscript readings (given in the footnotes of the printed edition) with his own readings. When manuscript readings seemed relevant, especially when all manuscripts had the same readings, they were substituted of those of the main text and translated. This is always indicated in a footnote. Some alternative readings, when they were thought meaningless, were not taken into account.

2   Technical Translations

Sanskrit words of the commentary have been coded, as much as it was possible, with standard English translations. This is especially the case of the mathematical vocabulary: we have tried to keep for each Sanskrit word a single English word to translate it. The idea was to see if differentiated uses of synonyms could hence be brought to light. For instance the Sanskrit word *karaṇa* has always been translated as "procedure". And Sanskrit words with similar meanings such as *upāya, ānaya* have been translated with different English words. The first occurrence of "procedure" in the translation is followed by a Sanskrit transliteration of *karaṇa* between parentheses. After that when the word "procedure" is found in the translation, one can be assured that *karaṇa* is the Sanskrit word it translates, unless another word is specified in between parentheses.

Sometimes, the commentator discusses terminological and grammatical problems, where the standard terminology makes the argumentation obscure. In these cases, we have reverted to a literal translation, recalling the coded one in between parentheses. Both literal and standard translations are given in the Glossary. From time to time, the unusual use of a word, verb, or technical grammatical terms which have not been entered in the Glossary has been indicated by a transliteration in between parentheses.

3   Compounds

When Bhāskara analyzes a compound in Sanskrit, it is done in a very standard way, so that its literal meaning and the type of compound it is are understood simultaneously. This is difficult to translate into English. I have attempted to do so and indicated the nature of the compound in between parentheses. These compound names (*dvandva, tatpuruṣa, bahuvrīhi, . . .*) as such are not stated in the Sanskrit text, but they are comonly used by Indologists. For instance, in verse 8, Āryabhaṭa uses a compound to indicate an operation involving two numbers. This is stated in Sanskrit, in the following declinated compound: *āyāmaguṇe*. This is how we have translated Bhāskara's analysis of the compound: "Those two which are multiplied by the height are these *āyāmaguṇe* (a *bahuvrīhi* in the dual case)." Technical compounds, when they are discussed grammatically, are always given a literal translation, which might differ from the coded, standard one we have adopted, however because such compounds will be the object of a discussion which will explain why a technical understanding should be used, there will be no confusion.

4   Numbers

In Sanskrit some numbers may be expressed by simple computations as in French "quatre-vingt" and "soixante-dix", even though they may also have a simple name of their own. They may also be expressed by metaphors, as *śaśin* (the moon) to express "one". When these forms are used commonly, we will translate the numbers by their common name in English, and add the Sanskrit transliteration in between parentheses. Higher numbers are sometimes expressed by enumerative *dvandvas* giving the digits in increasing order of their powers of ten, from left to right. This is the reverse order of the way they are noted. In this case we have respected the order of the compound, and presented in the form of a compound of digits with English names, the number named in this fashion. For example, in example 1 of BAB.2.4. the number 625, is expressed in the verse by the compound *śara-yama-rasa* that has been translated in the following way:

> five (*śara*)-two (*yama*)-six (*rasa*)

Mythological names of numbers can be found in section 4 of the Glossary.

5 Synonyms

Certain words are given by Bhāskara as synonyms. This means that they are part of a list, ending with the expression *ity paryāyāḥ* (meaning: those are synonyms). However they are not to be considered as absolute synonyms. Although this is not stated in the commentary itself, it appears quite obvious as one examines their respective uses in the treatise and the polysemy of Sanskrit words, that they should always be considered as local synonyms[140]. In the Glossary, such words have been noted as "synonyms" of the word commented. Words that simply gloss another word by apposition or substitutions are not indicated in the Glossary. I have inasmuch as it was possible, tried to adopt a different translation for each word. When this was not possible, the substitution of one word by another is indicated in parentheses.

6 Paragraphs

We have not followed closely the paragraph sectioning of Shukla's critical edition. This sectioning of the text is arbitrary: in manuscripts it does not appear and the printed edition openly introduced its own segmentation of the text[141]. In addition to those of the printed edition, other paragraphs have been introduced in order to separate clearly new developments in the commentary.

Where staged discussions occur, we have used the following rule:

---

[140]I do not know of any study on the notion of synonym as practiced in Sanskrit commentaries. That meanings are local may be underlined by the commentator as he uses the demonstrative *tad* (that) when concluding the analysis of a word or compound. This seems to mean: this is the local meaning of the given word. For a very brief account of Sanskrit grammarian's reflections on synonymy and polysemy, see [Potter 1990; pp. 7-8; 81-82].

[141][Shukla 1976; Introduction 10.2.iii, p. cxv].

When an alternative interpretation or a detailed discussion is given, it is separated from Bhāskara's answers, and indicated by the label: ⟨Objection⟩ or ⟨Question⟩ according to the nature of the statement.

When questions concerning the procedure described in a verse underline the way the commentator glosses one by one every word of the verse, the questions have been separated by Bhāskara's answer but bear no label.

When questions concern a point of detail of a grammatical analysis or of a given development not bearing any specific relation to the whole development, they have not been separated typographically from the rest of Bhāskara's commentary.

7   Examples

Some examples are versified, others not. The versified examples in Shukla's edition are in bold characters, the non-versified in plain font. However, in this translation, all examples will be in bold in order to stress visually the structure of the commentary. Examples have been numbered by the editor. Although these numbers, in the Sanskrit edition, are stated at the end of each versified example, in this translation they are given in the beginning. It seemed more natural to the English format.

The Translation

Chapter on Mathematics

[Benediction]

1. Homage to that Śiva[142] whose name, only when meditated upon, creates and destroys ⟨respectively⟩ good fortune and misery for gods, demons and men|

5. Whose pair of lotus like feet are rubbed by the foreheads of Kṛṣṇa and the Lotus-Born *(Brahmā)*‖

2. master[143] Āryabhaṭa, having propitiated the Lotus-Born by his stainless austerities obtained the true seed (*bīja*) of great importance whose scope is the essence of the motions of planets (*grahacāra*)|

On this, which is beyond the range of the senses, which has a clever meaning which is a clear, broad and concrete subject, a bit of commentary (*vyākhyāṇa*), obtained at the feet of a *guru*, is now written down by me‖

[Specifying the subjects to be treated]

Now three topics escaped from master Āryabhaṭa's lotus mouth: mathematics (*gaṇita*), time-reckoning (*kālakriyā*), and the sphere (*gola*). That which is this mathematics is two-fold and enters in four. Two fold, that is increase (*vṛddhi*) and decrease (*apacaya*). Increase is addition (*saṃyoga*); decrease is subtraction (*hrāsa*). Mathematics as a whole (*aśeṣagaṇita*) is covered by means of these two parts. And ⟨this⟩ is stated:

Varieties of addition are multiplication (*guṇanā*) and exponention (*gata*, which is raising to a power), and, ⟨varieties of⟩ subtraction are told to be division and roots of exponention (e.g. root extraction)[144].|

---

[142]Please refer to the section of the Glossary on Gods and Mythological Figures.

[143]This is how we have translated here the Sanskrit *ācārya* which is used for a venerated master.

[144]The expression refers to the operation of extraction rather than to its result, the root (*mūla*).

Having seen ⟨that it⟩ is covered by increase (*upacaya*) and decrease (*kṣaya*), one should know that this discipline (*śāstra*) is made of only two|

Varieties of addition, which ⟨are varieties⟩ of increase, are multiplication and exponention; and these are ⟨as follows⟩:

Guṇanā is a product (*abhyāsa*) of two different quantities, as twenty is ⟨the product⟩ of four and five. *Gata* is a product of same quantities ⟨as in⟩ squaring and cubing. A double-*gata* (*dvigata*) is a square, as sixteen is ⟨the product of⟩ four and four. Likewise a triple-*gata* (*trigata*) is a cube, as sixty four is ⟨the product of⟩ four and four and four.

p.44,
line 1

⟨As for⟩ "and ⟨varieties of⟩ subtraction". In this case the word "and" (*ca*) is read for the purpose of union. Therefore, indeterminate increase[145] in series, pulverisers, etc., and in the world, is included.

⟨As for⟩ "and ⟨varieties of⟩ subtraction are division and evolution". Varieties of subtraction, which ⟨are varieties⟩ of decrease, are division and roots of exponentions . In this case also, indeterminate decrease in series, pulverisers, and so on, and in the world, is included, on account of the word "and". It is in this way in the treatise and in the world; there is no such kind of mathematics which is not made of increase or[146] of decrease.

5

⟨Objection⟩

If it is so, in what way is the calculation (*prakriyā*) understood in this case: When one fourth is multiplied by one fifth one twentieth is produced. And this multiplication is said to be a variety of increase. However it unexpectedly turns out as (*āpatita*) a variety of decrease . When ⟨one performs⟩ the division of one twentieth by one fourth, then one fifth is seen. In this way, this which is ⟨described as⟩ a variety of decrease unexpectedly turns out as a variety of increase.

10

In both cases a refutation (*parihāra*) is told ⟨as follows⟩: In a rectangular field (*āyatacaturaśrakṣetra*) of four by five there are twenty quadrilateral fields (*caturaśrakṣetra*; in this case one should understand: squares)[147]. In that case, the length

---

This formulation is surprising and not used in other parts of the commentary. Below, while commenting on "and ⟨varieties of⟩ subtraction", Bhāskara uses the expression: *gatanām mūlāni*, "roots of *gatas*". We have translated this expression by "roots of exponentions". In verse 4, concerned with root extractions, and thereafter, this expression is not used anymore. The author only mentions the result of the procedure, the root itself.

[145] Reading as in Mss E. "*aniyatasvarūpā vṛddhiḥ*", rather than "*aniyatasvarūpavṛddhiḥ*", as in the main text of the printed edition.

[146] The Sanskrit word for "or", *vā*, is non-exclusive [Renou 1984; §382.B], so that all computations in "these mathematics" should be understood as being made of only increase, only decrease, or of both.

[147] Reading as in all manuscripts:

āyatacaturaśrakṣetre catuḥpañcake viṃśaticaturaśrakṣetrāṇi

Rather than the text inserted by the editor.

of one ⟨small square⟩ is one fifth ⟨of the length of the big rectangle⟩ and the breadth is one fourth ⟨of the breadth of the big rectangle⟩. The product of the two is the area (*phala*) ⟨of one small square⟩, one twentieth of the ⟨area of the rectangular⟩ field. "The division of one twentieth by one fourth" is therefore not a mistake (i.e. it appears as increase but really it is a geometrical decrease)[148].

This is a refutation in the mathematics of fields (*kṣetragaṇita*). An attempt should be made aiming at a refutation in the mathematics of quantities (*rāśigaṇita*).

Another says- "mathematics is two fold: quantity (*rā'si*) and field (*kṣetra*)[149]'. As in the *karaṇī* operation (*parikarman*):

Because it produces (*karoti*) the equality of the hypotenuse (*karṇa*) to the sides (*bhuja*) it is therefore ⟨called⟩ a *karaṇī*[150]|

Mathematics (*gaṇita*) is of two kinds: Mathematics of fields and mathematics of quantities. Proportions (*anupāta*), pulverisers, and so on, which are specific ⟨subjects⟩ of mathematics (*gaṇitaviśeṣa*), are mentioned in the mathematics of quantities; series, shadows, and so on, ⟨are mentioned⟩ in the mathematics of fields. Therefore, in this way, mathematics as a whole rests upon the mathematics of quantities or the mathematics of fields. That which is this *karaṇī* operation is only in the mathematics of fields. And even if in another case (i.e. not concerning the mathematics of fields) the *karaṇī* operation ⟨is performed⟩ and it does not explain (*pratipādakatva*) the three sides of a right-triangle (*karṇabhujākoṭi*) there is no mistake ⟨because truly⟩ that which is this *karaṇī* operation explains the hypotenuse etc.

⟨As for: mathematics⟩ "enters in four", four seeds, it enters in these.

Mathematics has been stated. Time-reckoning and the Sphere will be mentioned here and there.

---

[148]Reading:

> *viṃśatibhāgasya caturbhāgena bhāgaḥ iti na doṣaḥ*

rather than the printed edition:

> *viṃśatibhāgasya caturbhāgaḥ iti na doṣaḥ*
> "One twentieth of one fourth" is not a mistake

Which does not make much sense.

[149]Reading Shukla's emendation of the text, rather than the *kālakṣetra* of all manuscripts.

[150]The compound *karṇabhujayoḥ* is nonsensical grammatically. If one reads *karṇam bhujayoḥ* it is close to the initial reading but not the construction of a sentence with *samatva*. If one reads *bhujayoḥ karṇasya* it makes sense but far from the manuscript readings. As this verse appears to be a *gathā*, that is one whose quarters each follow a different pattern, no metric consideration can help us restore the original form of the compound. However the meaning of the verse in itself is quite clear.This verse is translated and analyzed in [Hayashi 1995; I.6.2, p.62], we discuss it also briefly in the Introduction.

Here as master Āryabhaṭa is starting the treatise (*śāstra*), a salutation to a favorite god, in ⟨his⟩ mind, is urged by devotion indeed[151] :

p.45,
line 5

Ab.2.1. Having paid homage to Brahmā, Earth, Moon, Mercury, Venus, Sun, Mars, Jupiter, Saturn, and the Group of Stars|
Here Āryabhaṭa states the knowledge honored in Kusumapura‖

brahmakuśaśibudhabhṛguravikujagurukoṇabhagaṇān namaskṛtya|
Āryabhaṭas tv iha nigadati kusumapure 'bhyarcitam jñānam‖

"**Brahmā**" is his favorite ⟨god⟩. Because a salutation to a favorite god urged by devotion destroys the obstacles that interrupt the work one craves and desires ⟨to finish, he starts by paying homage to him⟩. 10

Or else, Brahmā is the most distinguished of all gods because his feet are adorned by the garland of rays of the gems in the crowns of gods and demons; therefore the master, at the beginning, has paid homage to him.

Or else, a composition briefly ⟨stating⟩ the topics of the *Svāyaṃbhuva Siddhānta* has been undertook by the master, and the father of the *Svāyaṃbhuva Siddhānta* is the honored Creator (Brahmā)[152]. Therefore it is proper for him (Āryabhaṭa) to first make a salutation to him (Brahmā).

The motion of planets depends on latitudes (*akṣa*) and longitudes (*deśāntara*) and 15
these two specific longitudes and latitudes are due to the Earth[153], therefore after him (Brahmā), salutation is made to **Earth**.

Salutation has been made to the moon, and so on which stand each one higher than the preceding one, because this treatise concerns their motion.

Brahmakuśaśibudhabhṛguravikujagurukoṇabhagaṇa is Brahmā and Earth (*ku*) and Moon (*śaśin*) and Mercury (*budha*) and Venus (*bhṛgu*) and Sun (*ravi*) and Mars (*kuja*) and Jupiter (*guru*) and Saturn (*koṇa*) and Group of Stars (*bhagaṇa*) (a *dvandva* compound)[154]. Therefore the meaning is "**having paid homage to**" (*namaskṛtya*), that is having saluted (*praṇamya*), those that are "**Brahmā, Earth, Moon, Mercury, Venus, Sun, Mars, Jupiter, Saturn, and the Group of Stars**". Stars (*bha*) 20
are luminaries (*jyotis*) as Aśvinī and so on ; a group (*gaṇa*) of these is a group of stars (*bhagaṇa*; a *tatpuruṣa*).

---

[151]This sentence alone is nonsensical, it seems to be the fusion of two portions of fragments of sentences.

[152]Svayaṃbhū (as indicated in the section of the Glossary on the names of gods and planets) is one name of Brahmā. *Siddhānta* is the name of one genre of astronomical treatise, see [Pingree 1981; p.17sqq].

[153]Motions of planets are observed from the earth and corrections are necessary according to the longitude and latitude of the observer this may be why Bhāskara associates them to the earth.

[154]See section 3 called 'On the translation' for our dealings with Sanskrit compounds.

What should be told here concerning the higher and higher position of the moon and so on, this we will tell in the chapter on time reckoning.

⟨As for⟩ "**Āryabhaṭa**", by mentioning his own name, he indicates that there are other[155] works made according to the *Svāyaṃbhuva Siddhānta*; therefore, since there are a great number of works made according to the *Svāyaṃbhuva Siddhānta*, it would not be known by whom this work was composed. This is why he mentions his name. As ⟨in the *Arthaśāstra*[156]⟩: "This treatise has been made by Kauṭilya." [*Arthaśāstra*, 1.1.19]

The word "*tu*" fills the quarter of verse. [The word "*iha*"] indicates his city.

Nigadati is "**states**" (*bravīti*).

⟨As for⟩ "**the knowledge honored in Kusumapura**". Kusumapura is Pāṭaliputra[157]; he states the knowledge honored there.

This has been handed down by tradition: this *Svāyaṃbhuva Siddhānta*, as we traditionally know, has been honored by the experts who live in Kusumapura, even though there are ⟨other *siddhāntas*⟩ the *Pauliśa, Romaka, Vāsiṣṭha* and *Saurya*[158]. Because of this he asserts "The knowledge honored in Kusumapura".

[Assigning places to numbers]

In order to assign places (*sthāna*) to numbers (*saṅkhyā*), he states:

Ab.2.2. One and ten and a hundred
And one thousand, now ten thousand and a hundred thousand, in the
      same way a million|
Ten million, a hundred million, and a thousand million.
A place should be ten times the ⟨previous⟩ place‖
      ekaṃ ca daśa ca śatam ca sahasram tv ayutaniyute tathā prayutam|
      koṭyarbudaṃ ca vṛndaṃ sthānāt sthānam daśaguṇam syāt‖

One sets forth the places of numbers for the sake of easiness. For, if not, mathematical operations (*gaṇitavidhi*) would be difficult because of the lack of any

---

[155]Reading as in all manuscripts *svasaṃjñābhidhanenāsyāḥ* rather than *svasaṃjñābhidhanenānyāḥ*as in the printed edition.

[156]The *Arthaśāstra* of Kauṭilya is one of the classics of Sanskrit literature. It is a treatise on politics. For another discussion on the fact that Āryabhaṭa states his name, see BAB.1.1. in [Shukla 1976, p.5].

[157]This is the name of a city, ancient capital of the Mauryan empire and famous learning center. See [Sharma & Shukla 1976; intro, p.xvii].

[158]These are the name of four of the five *siddhāntas* summed up in Varāhamihira's *Pañcasiddhāntika*. The Paitamahāsiddhānta, the fifth siddhānta of Varahamihira's treatise, being inspired by the *Brahmasiddhānta*. See [Neugebauer & Pingree 1971].

assignement of places to numbers. Why? When placing the abundance of units (*rūpa*), many units have to be placed[159]. On the other hand, truly, when the places are settled, that operation (*karman*) to be accomplished with many units, can be performed with a single one only.[160]

⟨As for:⟩ "**One and ten and a hundred and a thousand**". One, ten, a hundred and a thousand have the first, second, third, and fourth place. "*Tu*" (**now**) is an expletive. *Ayutaniyute* is **ten thousand** (*ayuta*) and **hundred thousand** (*niyuta*; *ayutaniyute* is a *dvandva*). *Ayuta* has the fifth place. *Ayuta* is ten thousand. *Niyuta* has the sixth place. *Niyuta* is a *lakṣa* (a hundred thousand). "**In the same way**" (*tathā*), that is, in exactly the same manner, *prayuta* has the seventh place. Ten *lakṣas* are a *prayuta* (**a million**). *Koṭi* has the eighth place. A hundred *lakṣas* are a *koṭi* (**ten million**). *Arbuda* has the ninth place. Ten *koṭis* are an *arbuda* (**a hundred million**). *Vṛnda* has the tenth place . A hundred *koṭis* are a *vṛnda* (**a thousand million**).

⟨As for⟩ "**A place should be ten times the ⟨previous⟩ place**".   Another place ⟨should be⟩ ten times the ⟨previous⟩ place; it amounts to: the next place is ten times one's own ⟨previously⟩ singled out place.

⟨Objection⟩

For what purpose is this ⟨fourth quarter of verse⟩ stated? For certainly these places ⟨given in the first three quarters⟩ are ten times ⟨in worth⟩ in regard to the immediately adjoining ones. If the statement ⟨in the fourth quarter is made⟩ in order to understand places other than those ⟨given in the first three quarters⟩ (that is, the places coming afterwards) then the naming of places is useless .

Why?

⟨Objection⟩

By means of this very ⟨statement⟩, "a place should be ten times the ⟨previous⟩ place", ⟨the values of the places for numbers⟩ are named[161], because an understanding of the named places is established[162].

---

[159]This somewhat tautological sentence, may be easier to understand if we consider that *rūpa* here means *simultaneously* a "unit" and a "sign". In other words, when a big number is to be disposed on a working surface (e.g. "placing the abundance of units") if there is no positional notation, then it will be set down with a great number of symbols which together will amount to that big number (e.g. "many units/signs will have to be placed"). Please see the Glossary for the different meanings endorsed by *rūpa* in Bhāskara's commentary.

[160]This paragraph could refer to one or several non-positional notational systems. As in the Roman notational system for instance, one would consider the sum of the values of all the written symbols. However such a system is not known of in India.

[161]Because *abhihita* is in the nominative feminine, one should probably understand "the numbers are named", implicitly understanding "the values of the places of numbers".

[162]In other words, just this part of the rule is sufficient to know how the places should be named, since their value is known.

This is not a drawback. "A place should be ten times the place ⟨immediately before it⟩", this ⟨statement⟩ is a characterization (*lakṣaṇa*). The ⟨names of⟩ places beginning with one are illustrations (*udāhṛta*) of this characterization.

⟨Objection⟩

This is not so. For *sūtra* makers who wish to express ⟨statements⟩ concisely will not tell ⟨both⟩ characterizations ⟨and⟩ illustrations (*udāharaṇa*).

⟨Bhāskara⟩

It should not be understood in this manner. Since a characterization and an illustration are nonsensical ⟨here⟩, then the names of numbers[163], beginning with one and ending with a thousand million, are fixed ⟨by the first three quarters⟩. ⟨With the fourth quarter⟩ "a place should be ten times the place ⟨immediately before it⟩", merely an assignment of places to a number beginning with one, is indicated, ⟨and⟩ not the names of numbers, because ⟨the rule⟩ is of no use ⟨for naming numbers⟩[164].

Here this may be asked: What is the power (*śakti*) of the places, ⟨that power with⟩ which one unit becomes ten, a hundred, and a thousand? And truly if this power of places existed, purchasers would have shares in especially desired commodities. And according to ⟨their⟩ wish what is purchased would be abundant or scarce.[165] And if this was so, there would be the unexpected possibility for things to be different in worldly affairs (*lokavyavahāra*).

This is not wrong. Units configured (*vyavasthita*) in a place are made into ten, etc.

What then ⟨can be done⟩ with these ⟨units⟩?

These are explained by a traditional rule of writing (*lekhāgama*), or we have said ⟨previously⟩ that places were undertaken for the sake of easiness. And the setting down of places is:

○ ○ ○ ○ ○ ○ ○ ○ ○○

---

[163]Understanding the singular genitive here as referring to generality, rather than to a single number which would then be 111111111.

[164]All this paragraph is difficult to read. For instance, the last sentence of this paragraph seems to contradict the previous statement that the first part of the verse is concerned with the names of numbers):

> *upayogābhāvānna saṅkhyāsaṃjñā*
> lit. Not names of number because of no use.

This is however what we understand: the three first quarters of Ab.2.2 fix the names of number up to 10^9. [Hayashi 1995] has shown that higher numbers didn't have fixed names. However, this is not a rule explaining how to make the name of numbers. In addition it also gives a name to the first nine places of the decimal-place value notation. These places are defined as representing an increasing set of multiple of tens.

[165]In other words, if the place decided the value, in the world as well as in the treatise, one could buy a small amount, and then increase it afterwards, by simply changing its place.

[Operations on squares]

He states the first half of an *āryā* to expose operations on squares (*vargaparikarman*):

> **Ab.2.3.ab. A square is an equi-quadrilateral, and the area/result (*phala*) is the product of two identicals|[166]**
>
> **vargaḥ samacaturaśraḥ phalaṃ ca sadṛśadvayasya saṃvargaḥ|**

Varga, karaṇī, kṛti, vargaṇā, yāvakaraṇa are synonyms.

That which has four equal sides is this **equi-quadrilateral** (*samacaturaśra* is a *bahuvrīhi*), a specific field; that is **a square**.

The specific equi-quadrilateral field is the named (*saṃjñin*), square is the name (*saṃjñā*). In this case, because the named and the name are not distinguished, it is stated, figuratively: "A square is an equi-quadrilateral". As ⟨in the expression⟩ "Devadatta is a lump of flesh". Then[167], in this case, in so far as it is a specific equi-quadrilateral field, there is a possibility for the name "square" to be ⟨given to⟩ all those ⟨fields⟩, even when ⟨they are⟩ undesirable.

⟨Question⟩

In which other cases is there a possibility for the name "square" to be ⟨given to⟩ an undesirable specific equi-quadrilateral field?

It is replied: This kind of equi-quadrilateral with unequal diagonals (*asamakarṇa*) would have ⟨that name⟩ (Figure 1), and this ⟨field made of⟩ two equi-trilateral fields (*dvisamatryaśrakṣetra*) placed as if upraised, would have ⟨that name⟩ (Figure 2).

---

[166]One can understand the verse as meaning:

> A square is an equi-quadrilateral and the result which is the product of two identical ⟨quantities⟩ |

or

> A square is an equi-quadrilateral and ⟨its⟩ area is the product of two identical ⟨sides⟩ |

It is probably ambiguous in order to collect all these significations. Previous translators of this verse have noted this ambiguity. See [Sengupta 1927; p.13], [Clark 1930; p.21], [Shukla 1976; p.34]. Bhāskara expounds the verse in both directions.

[167]Reading *tathātra* rather than the *anyathātra* of the printed edition and the *yathātra* of four manuscripts. The opposition suggested by *anyathā* does not have much meaning here.

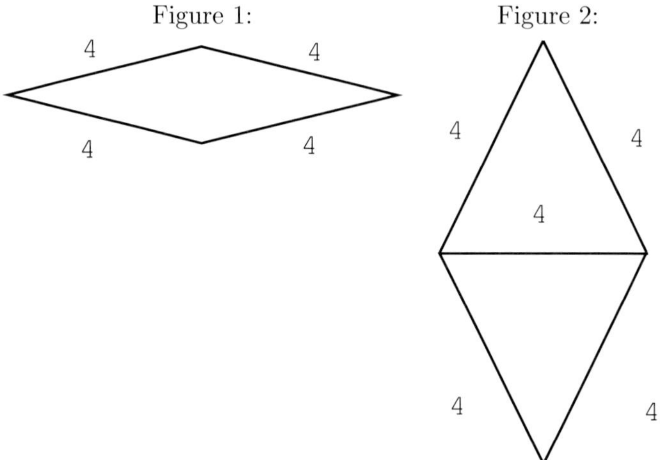

⟨Question⟩

What is wrong with the possibility for the name "square" ⟨to be given to these fields⟩?

It is stated: "And ⟨its⟩ area (*phala*) is the product of two identical ⟨sides⟩", therefore, the product of two identical ⟨sides⟩ should give the area, and such is not as wished in the above cases.

⟨Question⟩

In which case ⟨is it right for the name "square" to be given to a specific equi-quadrilateral⟩?

One should mention the diagonals (*karṇa*). Thus, a square is a specific equi-diagonal-equi-quadrilateral field. Or perhaps, one intends to know the name "square" only for a kind of equi-quadrilateral field secondarily characterized (*upalakṣita*) by two diagonals which have the same values (*saṅkhyā*).

⟨Question⟩

Why?

Because a discipline is not practiced in order to reach an unintended ⟨goal⟩. Or perhaps, in the world the name "equi-quadrilateral" has not at all been correctly established for an equi-quadrilateral field particularized by such shapes ⟨as in Figures 1 and 2⟩.

⟨Objection⟩

Because a square operation (*vargakarman*) exists in rectangular fields (*āyata-caturaśrakṣetra*), and so on, there is the possibility for the name "square" to be ⟨given to⟩ fields which are not equi-quadrilaterals (*asamacaturaśra*) also [168].

---

[168] What is exactly called here a *vargakarman* remains ambiguous. Please refer to volume II, section A.1 on page 2.

This is not a mistake. In these ⟨fields⟩ too, a square is the area of an equi-quadrilateral field. It is as ⟨in the following case⟩:

When one has sketched an equi-quadrilateral field and divided ⟨it⟩ in eight[169], one should form four rectangles whose breadth (*vistāra*) and length (*āyāma*) are three and four and whose diagonals are five. There, in this way, stands in the middle an equi-quadrilateral field whose sides are the diagonals of the ⟨four⟩ rectangles which were the selected quadrilaterals. And the square of the diagonal of a rectangular field[170] there, is the area in the interior equi-quadrilateral field[171]. Just this exposition (*darśana*[172]) ⟨exists⟩ in a trilateral (*tribhuja*) also, because a trilateral is half a rectangle. And a field is sketched in order to convince the dull-minded:

Figure 3:

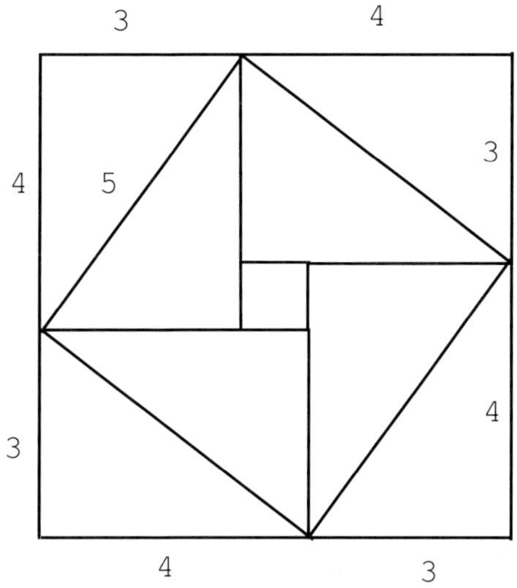

---

[169]The status of the expression *aṣṭadhā* (eight-fold) is difficult to understand here. Does it refer to the length of the sketched square (this seems to be the opinion of the editor, as seen on the printed figure p. 48), or is it a set expression concerning the following sketching of the rectangles (as suggested by Takao Hayashi and as it can be seen in manuscripts)? This is discussed in the Supplement for BAB.2.3., Volume II, paragraph A.

[170]The word *āyata* in the compound *...vargāyatakarṇa* is nonsensical. We have thus not read it.

[171]The use of the locative form here suggests that the area is not just the measure of a surface but simultaneously the surface itself delimited by the sides of the square.

[172]Another interpretation of the word *darśana* could be "display", and would therefore be referring to the drawn figure itself; this sentence would then be translated as:

> This very display ⟨is used⟩ in a trilateral also, because a trilateral is half a rectangle.

However, this is the only occurrence of this understanding of the word *darśana* with this meaning in Bhāskara's text.

p.49,
line 1

Therefore, every single square is a specific equi-quadrilateral.

⟨As for⟩ **"and ⟨its⟩ area is the product of two identical ⟨sides⟩"**. "*saṃvarga*" (product) expresses the area (*kṣetraphala*) of this equi-quadrilateral. *Sadṛśadvaya* is a couple of identicals (a genitive *tatpuruṣa*). Or perhaps *sadṛśadvaya* is what is and two and identical (a *karmadhāraya*)⟨and therefore is⟩ the two same (*sama*), i.e. identical[173]. The product of two identicals (the compound is in the genitive case in the verse). *Saṃvarga, ghāta, guṇanā, hati, udvartanā* are synonyms.

⟨Objection⟩

The product of two identical ⟨sides⟩ is the area of this equi-quadrilateral. "The product of two identicals", the statement "intended side" (*iṣṭabāhu*) should be made here. For if not the product of any couple of identicals would give an area.

This is not so. For one who wants a ⟨particular⟩ area does not perform the product of two other ⟨values⟩ indicative of an other field. For one who wants cooked rice does not take dirt.

An example (*uddeśaka*):

> **1. The squares ⟨of the digits⟩ beginning with one and ending with nine,**
> **tell them one by one|**
> **And the square of a quarter of a hundred, and also ⟨the square⟩ of a**
> **hundred increased by that‖**

Setting down (Nyāsa): 1, 2, 3, 4, 5, 6, 7, 8, 9; a quarter of a hundred is 25, one hundred increased by that is 125.

The squares ⟨of the digits⟩ beginning with one and ending with nine are obtained, in due order, ⟨with the rule⟩ "and the result (*phala*) is the product of two identicals".

Setting down: 1, 4, 9, 16, 25, 36, 49, 64, 81.

The square numbers of the digits (*rūpa*) beginning with one and ending with nine, are to be stated by those whose[174] characterizations (*lakṣaṇa*) which are rules

---

[173]The sentence here is:

> *athavā sadṛśadvayaṃ ca taddvayaṃ ca samasadṛśadvayam*

It is difficult to give a precise meaning to it. Prof. J. Bronkorst has proposed the following reading:

> *athavā sadṛśa ca tad dvayaṃ ca samasadṛśadvayam*

which we have used for the translation. Manuscripts B and C omit a passage including this sentence. Manuscript D gives an even more corrupt reading of this part, and breaks off at the end of the sentence. This alternative reading is not given by the editor in footnotes. It reads:

> *athavā sadṛśadvayañ ca tad dvayañ ca samadṛśadvayam sadṛśa*

There is probably not much we can do, given our sources, to retrieve the exact sentence.

[174]Reading as in manuscripts D and E *evaṃ yeṣāṃ* rather than the *evameṣāṃ* of the printed edition.

(*sūtra*) are as follows:

**When one has made the square of the last term (*pada*), one should
multiply twice that very last term |
⟨separately⟩ by the remaining terms, shifting again and again (*utsārya
utsārya*), in the square operation‖**

Why? Because when the squares of the numbers ⟨beginning with one and ending
with nine⟩ are not known, then one cannot set down the square-number of the last
term. But for us, all is covered with just ⟨Āryabhaṭa's⟩ characterization[175].   20

The square of a quarter of a hundred is 625; ⟨the square⟩ of a hundred increased
by just this is 15625.   p.50,
line 1

The square of fractions (*bhinna*) is also just like this. However, when one has
made separately the squares of the numerator (*aṃśa*) and denominator (*cheda*)
quantities, that have been made into the same kind (*sadṛśa*), the result of the
division of the square of the numerator quantity by the denominator quantity is
the square of the fraction.

Example:

**2. Tell me the square of six and one fourth and of one increased by one
fifth |**   5
and of two minus one ninth ‖

Setting down: 6   1   2
         1   1   1
         4   5   9°

Procedure   (*karaṇa*): "⟨the whole number⟩ multiplied by the denominator and   10
increased by the numerator", therefore $\dfrac{25}{4}$ ⟨is obtained⟩.

Separately the square quantities of these denominator and numerator quantities
are 16, 625. When one has divided the square of the numerator quantity by the
square of the denominator quantity, the result (*labdha*) is $\dfrac{39}{1}$.
16

Likewise, ⟨the squares⟩ of the remaining ones also are, in due order, $\begin{matrix}1 & 3\\11 & 46\\25 & 81\end{matrix}$.

---

[175]In other words, Bhāskara does not endorse the rule he has quoted above.

[Operations on cubes]

He states the latter half of an *āryā* to expose operations on cubes (*ghanaparikar-man*):

15   **Ab.2.3cd. A cube is the product of a triple of identicals as well as a twelve edged ⟨solid⟩ ‖**

     sadṛśatrayasaṃvargo ghanas tathā dvādāśāśriḥ syāt‖

[The product of a triple of identicals] is *sadṛsyatrayasaṃvarga* (a genitive *tatpuruṣa*). A cube (*ghana*) is the product of a triple of identicals. *Ghana, vṛnda, sadṛśatrayābhyāsa* are synonyms. And this is a twelve edged ⟨solid⟩. That which has twelve edges is this twelve edged ⟨solid⟩ (*dvādaśāśri* is a *bahuvrīhi* compound). *Syāt* ("should be", the third singular optative form of the verbal root *as-*, meaning "to be") is *bhavet* (the third singular optative form of the verbal root *bhū-*, having the same meaning).

⟨Remark⟩

With the word *tathā* (as well as) he explains the equi-quadrilateralness (*sama-caturaśratā*) of a cube.

This is not so. Without the word *tathā* too one is just as able to know that this cube is equi-quadrilateral.

p.51, line 1

⟨Question⟩

Why ?

With the expression "product of a triple of identicals", ⟨Āryabhaṭa⟩ refers, for the surface (*phala*) of an equi-quadrilateral field, to a height (*ucchrāya*) which is equal to (*sadṛśa*) the side (*bāhu*) of that ⟨very equi-quadrilateral⟩ field, because the volume (*phala*) of a cube is the area of the field multiplied by the height.

Or perhaps, a mentioning of the equi-quadrilateral which was the subject (*adhikṛta*) of this ⟨expression (Ab.2.3.ab)⟩ "a square is an equi-quadrilateral" should be continued[176] ⟨in Ab.2.3.cd as well⟩. ⟨This cube's⟩ edges should be shown with clay or something else[177].

---

[176]The verbal root *anuvṛt* is a technical term used in commentaries to indicate that a phrase or a word of the previous verse should be supplied in the next one.

[177]This would be a reference to clay figures that would have been used in representing three-dimensional figures.

An example:

> **3. Tell me separately the cube of the digits (*rūpa*), beginning with one
>   and ending with nine |**
> **And also the cube of the square of eight times eight, and of the square
>   of the square of a quarter of a hundred ||**

Setting down:1, 2, 3, 4, 5, 6, 7, 8, 9 ; the square of eight times eight is 4096 ; the
square of the square of a quarter of a hundred is 390625.

The cubes of ⟨the digits⟩ beginning with one and ending with nine obtained, in
due order, with the expression "a cube is the product of a triple of identicals"   10
are 1, 8, 27, 64, 125, 216, 343, 512, 729.

In this case also, the cube-numbers of those ⟨digits⟩ beginning with one are to be
recited ⟨by those⟩ whose rule which is a characterization is "the cube of the last
place should be, etc."

⟨Question⟩

Why?

Because when the cube-numbers ⟨of one, etc.⟩ are unknown then indeed one is not
able to set down the cube-number of the last place.

[The cube] of the square of eight times eight is 68719476736, ⟨the cube⟩ of the
square of the square of a quarter of a hundred is also 59604644775390625.   15

The cube of a fraction is also just like that. An example:

> **4. Say, clearly, the cube-number of six, five, ten and eight that are
>   computed with as diminished by their respective parts|**
> **If ⟨you have⟩ a clear knowledge in cube-computations||**

Setting down:
$$\begin{array}{cccc} 5 & 4 & 9 & 7 \\ 5 & 4 & 9 & 7 \\ 6 & 5 & 10 & 8 \end{array}$$
                                                                                        20

The cubes obtained, according to ⟨the given⟩ numbers are: 25

$$\begin{array}{cccc} 198 & 110 & 970 & 488 \\ 107 & 74 & 299 & 191 \\ 216 & 125 & 1000 & 512 \end{array}$$

p.52,
line 1   [Square root]

In order to compute (*ānayana*) square roots (*vargamūla*), he states:

> **Ab.2.4. One should divide, constantly, the non-square ⟨place⟩ by twice
> the square-root|**
> **When the square has been subtracted from the square ⟨place⟩, the
> quotient is the root in a different place‖**

> **bhāgaṃ hared avargān nityaṃ dviguṇena vargamūlena|**
> **vargād varge śuddhe labdhaṃ sthānāntare mūlam‖**

5   *Bhāga* (division), *hṛti, bhajana* and *apavartana* are synonyms. One should take
away (*haret*) ⟨in other words⟩ one should seize (*gṛhṇīyāt*) that part (*bhāga*)[178].

Beginning with what place?

He says: "**From the non-square ⟨place⟩**" (*avarga*), what is not a square is a non-square; *from* that non-square (the expression is in the ablative case). In this computation (*gaṇita*), the square is the odd (*viṣama*) place. Since a non-square ⟨takes place⟩ when oddness is denied, by means of ⟨the affix⟩ *nañ*[179] ⟨the expression refers to⟩ an even (*sama*) place, because, indeed, a place is either odd or even[180].

By what should one divide?

He says: "**Constantly, by twice the square root**". That whose multiplier is two is this "two times" (*dviguṇa* is a *bahuvrīhi* compound). What is "that" ⟨whose multiplier is two⟩? The square-root (*vargamūla*). *By* twice the square root (the compound is in the instrumental case).

10   How, then, is this square-root obtained?

He says: "**When the square has been subtracted from the square ⟨place⟩, the quotient is the root in a different place**".

When the square has been subtracted from the square ⟨place⟩, that is, from the odd place; the meaning is: when a square-computation (*vargagaṇita*) ⟨is performed⟩. In this case, that quotient here becomes in a different place what is called the root.

A place other (*anyasthānam*) than a ⟨given⟩ place is a different place (*sthāna-antaram*); in this different place, this quotient has the name root. Where, however, a different place precisely does not exist, there that ⟨quotient⟩ has the name root in that very place ⟨where it was obtained⟩.[181]

---

[178]This is a gloss of the expression "to take away (or remove) the part" which is the Sanskrit way of expressing the operation of division.

[179]This is the name of the negation affix, here the *a* prefixed to *varga* in the word *avarga* (non-square).

[180]For an explanation of the mathematical content of this remark, see the Supplement for this verse commentary, Volume II, section B, p. 15.

[181]For an explanation of this paragraph, please refer to the supplement for this verse.

Why ?

Because a different place is not possible. This very rule is repeated again and again, until the mathematical operation (*gaṇitakarma*) is completed. 15

An example:

1. I wish to know, O friend, the root of the squares of the numbers seen formerly,|
That is ⟨the square-root⟩ of one and so on, and of the square quantity five (*śara*)[182]-two (*yama*)-six (*rasa*)‖

Setting down: 1, 4, 9, 16, 25, 36, 49, 64, 81, 625.

The square roots obtained in due order are : 1, 2, 3, 4, 5, 6, 7, 8, 9, 25. 20

An example concerning the computation (*ānayana*) of the roots of fractions (*bhinna*):

2. Having computed (*vigaṇaya*) in accordance with ⟨Ārya⟩bhaṭa's calculation (*saṅkhyā*)|
Tell the square roots of six with one fourth, and thirteen with four ninths ‖

$$\text{Setting down :} \quad \begin{array}{cc} 6 & 13 \\ 1 & 4 \\ 4 & 9 \end{array}$$

p.53, line 1

Procedure: When one has performed the product of the denominator (*cheda*) and the higher quantity (*uparirāśi*), one should add the numerator (*aṃśa*). What is produced is $\dfrac{25}{4} \ \Big| \ \dfrac{121}{9}$.

One by one, the roots of the numerator and denominator quantities are $\dfrac{5}{2} \ \Big| \ \dfrac{11}{3}$. 5

The result of the division of the root of the numerator quantity by the root of the denominator quantity is the square root of the fraction $1\dfrac{2}{2}$; and the square root of the fraction thirteen with four ninth is $2\dfrac{3}{3}$.

---

[182]Please see the Glossary of the Metaphoric and Peculiar Expressions to name numbers (Volume II, section 4, p. 2).

[Cube root]

In order to compute (*ānayana*) cube roots (*ghanamūla*), he says:

Ab.2.5. One should divide the second non-cube ⟨place⟩ by three times the square of the root of the cube|
The square ⟨of the quotient⟩ multiplied by three and the former ⟨quantity⟩ should be subtracted from the first ⟨non-cube place⟩ and the cube from the cube ⟨place⟩ ‖

aghanād bhajed dvitīyāt triguṇena ghanasya mūlavargeṇa|
vargas tripūrvaguṇitaḥ śodhyaḥ prathamād ghanaś ca ghanāt‖

What is not a cube is a non-cube, *from* that non-cube; [*Bhajed* (one should divide)], the meaning is: one should take away the part, one should seize the part.

Because there is more than one non-cube place, he states: "**from the second.**" In this computation, there is one cube, two non-cube ⟨places⟩.

Why is this ⟨said⟩: "there is one cube, two non-cube ⟨places⟩"?

It is replied: "the square ⟨of the quotient⟩ multiplied by three and the former ⟨quantity⟩ should be subtracted from the first non-cube", the first non-cube place is established (*siddhi*). ⟨It is stated⟩ "one should divide the second non-cube place", the second non-cube place is established. On the other hand, there is only one cube ⟨place⟩, because the second is not heard of.

Beginning with the second non-cube place, by what should one divide?

He says: "**by three times the square of the root of the cube**". That whose multiplier is three is this three times (*triguṇa* is a *bahuvrīhi* compound). ⟨Three times⟩ what? The square of the root of the cube. *By* three times that square of the root of the cube.

⟨As for:⟩ "**The square ⟨of the quotient⟩ multiplied by three and the former ⟨quantity⟩ should be subtracted from the first ⟨non-cube place⟩**". It is the square that is multiplied (*guṇita*) *by* three (*tri*) *and* by the former (*purva*) quantity ⟨this explains⟩ *tripūrvaguṇita* (an instrumental *tatpuruṣa* related to the noun *varga* (square) having *tripūrva* as a sub-dvandva, the word *rāśi* (quantity) being supplied to *pūrva*)[183]. The square of what? "Of the quotient" is the remaining part of the sentence.

Śodhya (should be subtracted) is *śodayitavya* (an obligation verbal adjective paraphrased by the causative form which has the same meaning). "From the first non-cube ⟨place⟩", is to be connected (*sambandhanīya*) ⟨to this verb⟩.

---

[183]The former quantity referred to here is the partial cube-root computed, please see the supplement describing the procedure in this verse (Volume II, section B, p. 15).

⟨As for:⟩ "**And the cube from the cube ⟨place⟩**". And the cube should be subtracted. From where? "From the cube". From the cube place. "Then the cube root is produced" should be supplied. In this case, when one has seen that this is the cube-quantity and has considered "one cube ⟨and⟩ two non-cube ⟨places⟩", from that place ⟨considered as the highest⟩ cube, one should make beforehand the cube-root, using that ⟨rule⟩ "the cube should be subtracted from the cube ⟨place⟩". Thereupon all of this *āryā* meter rule beginning with "One should divide from the second", has been examined (*upasthita*).

p.54,
line 1

An example :   5

**1. Tell me one by one the root of the cube quantities beginning with
one|
Let the cube root of eight (*vasu*)-two (*aśvin*)-seven (*muni*)-one (*indu*)
be computed quickly‖**

Setting down: 1, 8, 27, 64, 125, 216, 343, 512, 729, 1728.

Each cube root obtained is, in due order: 1, 2, 3, 4, 5, 6, 7, 8, 9, 12.

An example: 10

**2. Tell, correctly, according to ⟨Ārya⟩bhaṭa's treatise |
The root of the cube which amounts (*saṅkhyā*) to four (*kṛta*)-two (*yama*)-
eight (*vasu*)-nine (*randhra*)-six (*rasa*)-four (*abdhi*)-one (*rūpa*)-nine
(*randhra*)-two (*aśvin*)-eight (*nāga*).‖**

Setting down: 8291469824. The cube root obtained is 2024.

In exactly the same way, an example concerning the compution of the cube roots
of a fraction:

**3. Let the root , called fraction, of thirteen increased by what is called
three-zero (*śūnya*)-one (*rūpa*)|
Of which the denominator (*aṃśa*) is the cube of five be computed correctly in numbers‖**   15

Setting down: $\dfrac{13103}{125}$. The cube-root obtained is $2\dfrac{2}{5}$.

p.54,
line 20

[Area of a trilateral field]

Now, in order to compute (*ānayana*) the area of a trilateral field (*tribhujakṣetra*), he states:

Ab.2.6.ab The bulk of the area of a trilateral is the product of half the base and the perpendicular|[184]

tribhujasya phalaśarīraṃ samadalakoṭībhujārdhasaṃvargaḥ|

p.55,
line 1

That field which has three sides is this **trilateral** field (*tribhuja* is a *bahuvrīhi* compound to which the word "field" (*kṣetra*) may be supplied). *Bhujā*, *bāhu* and *pārśva* are synonyms. In this case there are three ⟨kinds of⟩ fields: equi⟨laterals⟩ (*sama*), isoceles (*dvisama*) and uneven ⟨trilaterals⟩ (*viṣama*). ⟨As for⟩ "**Of a trilateral**", that is: when one has accepted a category (*jāti*) of trilateral fields, it is indicated by ⟨using⟩ the singular voice[185]. Of a trilateral.

⟨As for,⟩ "**the bulk of the area**". The bulk (*phala*) of the area (*śarīra*) is *phalaśarīra* (a genitive *tatpuruṣa*); the meaning is: the size (*pramāṇa*) of the area.

⟨As for⟩ "**the product of half the base and the perpendicular**" . *Samadalakoṭī* is the perpendicular (*avalambaka*).

5   On this point some ⟨people⟩ explain :

---

[184]K. S. Shukla (in the line of Clark's interpretation [Clark 1930; p.36]) gives the following translation of this verse:

> "The product of the perpendicular (dropped from the vertex on the base) and half
> the base gives the measure of the area of a triangle" [Shukla 1976; p.38].

And P. C. Sengupta:

> "The area of a triangle is its *Sarira* (body) and is equal to half the product of the
> base and the altitude (...)" [Sengupta 1927; p.15].

The difference between these two translations lies, first of all, in the interpretation of the compound *phala-śarīra*.

For Sengupta it is a *karmadhāraya*, meaning "that which is the area which is the body." Shukla follows both Bhāskara's and Nīlakaṇṭha's interpretation of it as a genitive *tatpuruṣa*, meaning the "body/bulk of the area". The polysemy of the word *śarīra* may explain these different interpretations: *śarīra* may mean bulk, but also intrinsic nature, this explains Sengupta's reading of the verse.

The second reason for these differences arises from the translation of *samadalakoṭī*. According to commentators it is the height in a triangle. The problem is that this is not the compound's literal meaning (a mediator). Bhāskara comments on this point, below.

[185]As noted by J. Bronkhorst, this may be a recalling of the Pāṇinean sūtra 1.2.58:

> *jātyākhyāyām ekasmin bahuvacanam anyatarasyām*
> Plural optionally can be used for singular when *jāti* "class" is to be expressed.
> ([Sharma 1990; II. p. 129])

"That which has same parts is this *samadala* (a *bahuvrihī* compound). That which has same parts (*samadala*) and is an upright (*koṭi*) is this *samadalakoṭi* (a *karmadhāraya* compound)[186]."

For these ⟨people⟩ the area is established just for those trilateral fields (*tryaśrakṣetra*[187]) which are either equi-⟨laterals⟩ or isoceles, and not for an uneven-trilateral field.

For us, however, who are explaining ⟨the expression⟩ "*samadalakoṭi*" by that common acceptation (*vyutpatti*) ⟨of the word⟩ as "perpendicular" (*avalambaka*)[188], the computation (*ānayana*) of the area of all three ⟨types of fields⟩ is established. And also, for those who produce a grammatical derivation (*vyutpatti*) the computation of the area of the three trilateral fields is rightly established.

Why?

"In ⟨the case of⟩ conventional meanings, the action whose purpose is the performing of a ⟨grammatical⟩ derivation is not a purposeful action".[189]

Half (*ardha*) of the base (*bhujā*) is *bhujārdha* (a genitive *tatpuruṣa*). Now, in this case, although there is a possibility of accepting, with the word *bhujā*, the three sides (*pārśva*) in general (*sāmānyena*), since ⟨it is said that⟩ *bhujā*, *bāhu* and *pārśva* are ⟨synonyms⟩, only a specified side (*bhujā*) is chosen. It is called *bhujā*.[190] ⟨It is stated⟩:

"What impels generality lies in specificity".(*Mahābhaṣya* 4.1.3[191])

Here, in mathematics (*gaṇita*) the word *bhujā* is to be accepted as an *Uṇādi*[192].

For otherwise, ⟨according to Pāṇini's verse⟩

"⟨The words⟩ *bhujā*[193] and *nyubja* used in the meaning of arm and sickness ⟨are irregular forms⟩" *Aṣṭādhyāyī, 7. 3. 61.*

---

[186] *Koṭi* in a right triangle is the name of one perpendicular, the other being the base (*bhujā*) and the hypotenuse is called *karṇa*. So here *samadalakoṭi* which is a halver (*sama-dala*) and an upright side (*koṭi*) would be a "halving upright" or a "mediator".

[187] From here on, unless specified, this compound is used in the commentary to refer to a trilateral. The verse uses the compound *tribhuja*.

[188] From here on, unless stated, *samadalakoṭi* is the word translated as "perpendicular".

[189] In other words, a grammatical or etymological analysis of words with conventional meanings do not instruct us on their meanings.

[190] The idea is that *bhujā* meaning side is not just any side but a particular one, namely the base.

[191] See [Keilhorn ; II p. 246, line 6] This aphorism is also quoted in BAB.4.4, p.248, line 2. In other words, in this particular case, one gives a specific meaning to *bhujā* in order to give a general rule on the area of triangles.

[192] That is as a word whose etymology is not explained by Pāṇini's.

[193] In Shukla's edition the verse is given with *bhujā* (that is the word has a long "a") but in the editions of the Aṣṭādhyāyī the word is given with a short a. Whether this is a scribal error or a distortion by the commentator remains unclear.

Since the word *bhujā* with the meaning "arm" is an irregularly formed word ⟨according to Pāṇini⟩ it is not accepted in ⟨the sense of⟩ the side of a field ⟨and therefore should be listed as an *uṇādi*⟩. Half of this *bhujā* (meaning the base) is *bhujārdham*.

Samadalakoṭībhujārdhasamvargaḥ is the product of half the base with the perpendicular (a *tatpuruṣa*). It is the bulk of the area of a trilateral (*tribhuja*).

An example:

> **1. Friend, ⟨tell⟩ the areas of equi⟨laterals⟩ whose sides (*bhujā*) are ⟨respectively⟩ seven, eight, and nine|**
> **And of an isoceles whose base (*bhū*) is six, and ears (*śravaṇa*) five‖**

p.46, line 1   Setting down :

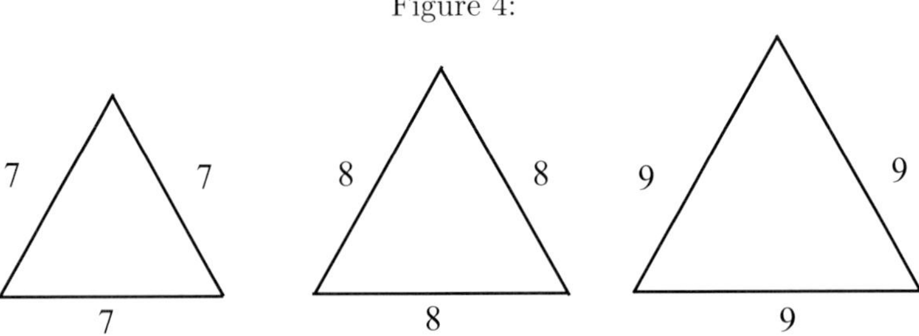

Figure 4:

These are three equi⟨laterals⟩.

For the isoceles also, the setting down is:

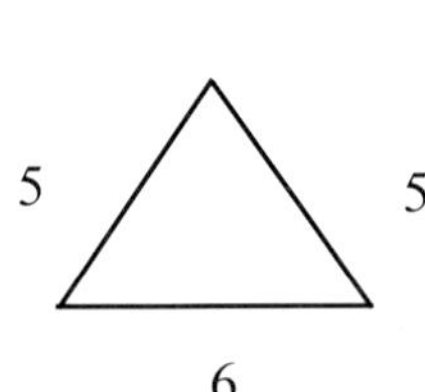

Figure 5:

Procedure:

> "In an equi-trilateral field the location of the perpendicular is precisely equal[194]."

The section of the base (*ābādhā* [195]) which is half of the base is $1\frac{3}{2}$.

"That which precisely is the square of the base (*bhujā*) and the square of the upright side (*koṭi*) is the square of the hypotenuse (*karṇa*)". [*Gaṇitapāda 17-a.*]

That is, the square of the hypotenuse is ⟨the sum of⟩ the squares of both the base and the height. Therefore, when the square of the base is subtracted from the square of the hypotenuse, the remainder is the square of the perpendicular, that is $3\frac{36}{4}$.

The perpendicular is $3\frac{36}{4}$ *karaṇīs*.

Half the base also is $1\frac{12}{4}$ *karaṇīs*. Therefore, since there is a product for two *karaṇīs*, the area of the field is obtained as "the product of half the side and the perpendicular", that is $3\frac{450}{16}$ *karaṇīs*.

In due order, exactly in the same way, the area of the two remaining equi⟨laterals⟩ are ⟨respectively⟩ [768 *karaṇīs*], and $3\frac{1230}{16}$ *karaṇīs*.

10

Since, for an isoceles trilateral also "The location of the perpendicular is precisely equal", a section of the base is 3. Using just the previous procedure, the perpendicular is 4. Using exactly the same procedure, the area is 12.

An example:

> **2. The two ears (*karṇas*) are indicated as being ten, and its base (*dhatrī*)**
> **is told to be sixteen|**
> **The calculation (*saṅkhyāna*) of the area of this isoceles should be told**
> **with caution‖**

---

[194]The use of *sama* (same) in this quotation is an elliptical way of expressing that the height sections the base in two *equal* segments.

[195]This is a technical term naming the two segments of the base delimited by the perpendicular.

15   Setting down:

Figure 6:

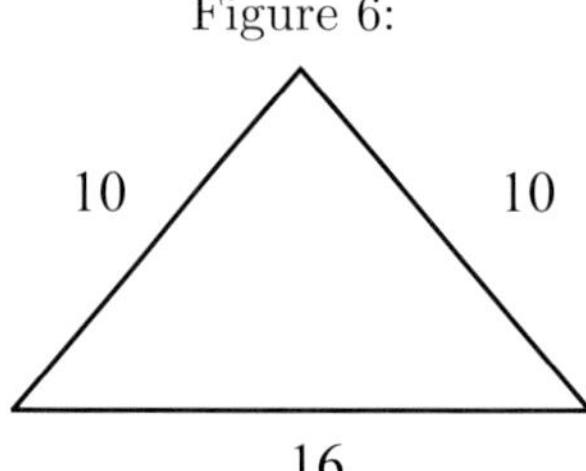

With the previous procedure, the area obtained is 48.

An example in uneven trilateral fields (*tribhujakṣetra*):

> **3. An ear should be thirteen, the other fifteen, the base (*mahī*) exactly**
>   **two ⟨times⟩ seven |**
> **Friend, what should be the value of the area of this uneven trilateral**
>   **field? ‖**

p.57,
line 1   Setting down:

Figure 7:

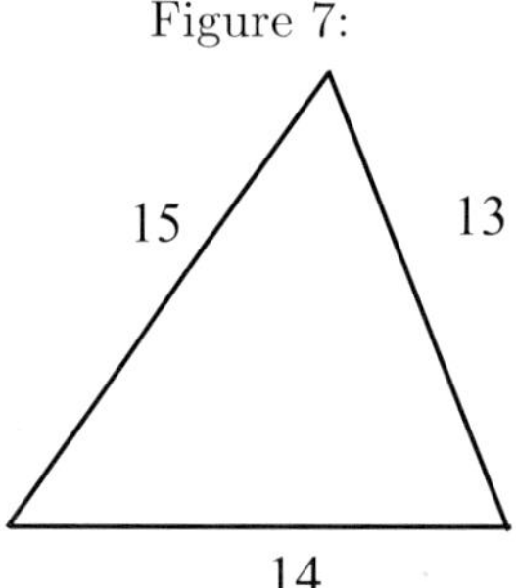

Procedure: In a trilateral field the difference of the squares of the two sides, or the product of the sum (*samāsa*) and the difference (*viśeṣa*) of the two, is the product of the sum and the difference of the different sections of the base (*ābādhāntara*).[196]

When one has divided by the base (*bhūmi*[197]) whose size is the sum of ⟨its⟩ different sections, a *saṃkramaṇa*[198] is ⟨applied⟩ to that very base together with the quotient (*labdha*).

---

[196] For an explanation of the computation described here and below, please refer to the supplement for BAB.2.6, Volume II, section C.1, p. 22.

[197] From here on, unless indicated, this is the word translated as "base".

[198] This is the name of a procedure given in Ab.2.24.

With this procedure (*krama*) the sizes of the two different sections of the base are obtained. With these two sizes of the different sections of the base, the computation of the perpendicular of an uneven trilateral field ⟨is performed⟩.

It is as follows: The square quantities of the two sides are 169, 225. Their difference is 56. The sum (*ekībhāva* [199]) of the sides is 28. Their difference is 2. Since the product of these two [is the differences of the squares of the sides] ⟨and is also equal to 56⟩, when this is divided by the base whose size, 14, is the sum of ⟨its⟩ different sections, what is obtained is 4.

Saṃkramaṇa ⟨is performed⟩ with this ⟨quotient⟩ together with the base (*bhū*); "increased or decreased by the difference" 18, 10 ⟨are obtained⟩. "Halved"; the different sections of the base are in due order 9, 5.

Using these two, the computation (*ānayana*) of the perpendicular of the trilateral field is ⟨as follows⟩: With the ear whose size is fifteen and with a different section of the base whose size is nine, the perpendicular (*avalambaka*) obtained is 12. With the ear whose size is thirteen and with a different section of the base whose size is five, the perpendicular obtained is again 12. The area is said to be: "the product of half the base (*bhujā*) and the perpendicular"; *bhujā* is the base, half of this is 7. The area brought forth, ⟨which is said to be⟩ "the product of half the base and the perpendicular", is 84.   15

An example:

4. The base should be fifty with one, an ear thirty increased by seven|
The other is said to be twenty; what the area of this uneven trilateral
field is, let it be told.‖

Setting down:

Figure 8:

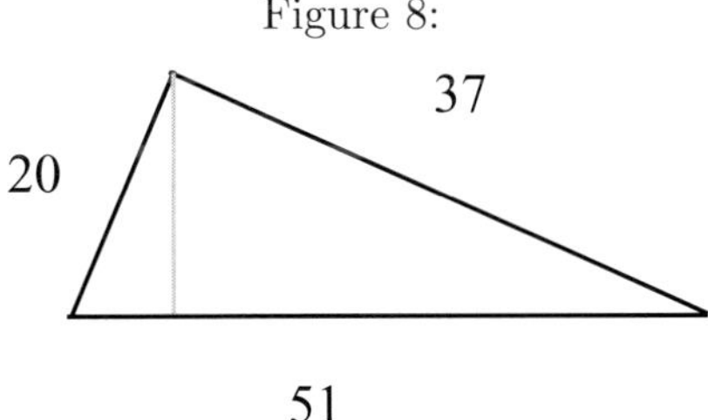

Litt.the state of becoming one.

And what is obtained with the previous procedure is the two different sections of the base, 16, 35; the perpendicular 12; the area 306.

[The volume of a six-edged solid]

In order to compute the volume (*ghanaphala*) of just that equi-trilateral field, he states the latter half of the *āryā*:

Ab.6.cd Half the product of that and the upward side, that is ⟨the volume of⟩ a solid called "six-edged"‖

ūrdhvabhujātatsaṃvargārdhaṃ sa ghanaḥ ṣaḍaśrir iti‖

The upward-side (*ūrdhvabhujā*) is a height (*ucchrāya*) in the middle of the field. "That" (*tat*) refers to the area (*kṣetraphala*). *Ūrdhvabhujātatsaṃvarga* is the product (*saṃvarga*) of the upward-side (*ūrdhvabhujā*) and that (*tat*) (a genitive *tatpuruṣa* with a sub-*dvandva*). *Ūrdhvabhujātatsaṃvargārdha* is half (*ardha*) of the product of that and the upward side (a genitive *tatpuruṣa*). ⟨As for:⟩ **"that is the solid"**, it amounts to : the volume (*ghanaphala*), and that ⟨solid⟩ is six-edged. That which has six edges is this six-edged ⟨solid⟩ (*ṣaḍaśri* is a *bahuvrīhi* compound)[200].

Now, when the size of the upward-side is known, one can state that the volume is half the product of that and the upward-side, but not when ⟨it is⟩ unknown.

This is true indeed. However, really, in this case, the size of the upward-side is known.

Why?

Because a method to compute (*ānayanopāya*) it is shown in ⟨another⟩ treatise.

It is as follows[201]: It is obvious (*pratyakṣa*) that the so-called "upward-side" is a height in the middle of the field. And that is the upright-side (*koṭi*) for the side (*bāhu*) of a *śṛṅgāṭaka* field which is located obliquely (*tiryak*) as an ear (*karṇa*), ⟨while⟩ the base (*bhujā*) is the intermediate space (*antārala*) in between the root (*mūla*) of the ear and the center (*kendra*) of the field.

When computing that ⟨base⟩, a Rule of Three: "If the side (*bāhu*) of a trilateral field is obtained with the perpendicular (*avalambaka*) of that very trilateral field,

---

[200]The word *ghana* of Āryabhaṭa's verse can be interpreted in two ways: as an abbreviated form of the compound *ghanaphala*, meaning "volume", and, as the noun to which the adjective "six-edged" (*ṣaḍaśri*) is related.

[201]Please refer to the supplement for BAB.2.6.cd (volume II, section C.2, p. 27) for an explanation of the following description and the reasonings exposed below.

then for the perpendicular whose amount is half the side of the ⟨initial⟩ trilateral field, how much is the side?"

The performance (*vidhāna*) of a Rule of Three with the hypotenuse (*karṇa*), base (*bhujā*) and upright-side (*koṭi*) of that ⟨trilateral⟩ has been established with other prescriptions and therefore is not stated here. And the prescription ⟨is two fold⟩: "That which is the square of the base and the square of the upright-side is the square of the hypotenuse" (Ab.2.17.ab.) and "That result quantity of the Rule of Three is multiplied, now, by the desired quantity." (Ab.2.26.ab.)   15

An example:

1. Tell me precisely and quickly, the computation of the volume of the
   śṛṅgāṭaka |
Whose sides are reckoned (*gaṇita*) as twelve, and the measure of its
   upward-side.‖

Setting down:

Figure 9:

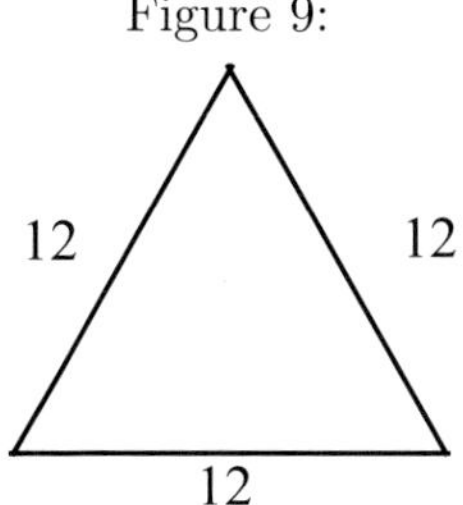

Procedure: "If an ear whose *karaṇīs* are a hundred increased by forty four is ob-   p.59,
tained with a [perpendicular] whose *karaṇīs* are a hundred increased by eight,   line 1
then with a perpendicular whose *karaṇīs* are thirty-six, how much is the ear
⟨obtained⟩?"

In order to show the proof (*upapatti*) of ⟨that⟩ Rule of Three, a field is set-down   5
and the Rule of Three is set-down: 108, 144, 36 [these are *karaṇīs*].

The inner ear obtained is 48 [*karaṇīs*]. This very ear is the [base] of the trilateral [field] located upwards (*ūrdhva*). The difference of the square (*kṛti*) of the
hypotenuse is the square of the upright-side. And that is 96. Here one should show
the upward-side with threads, sticks and so on.

The area of the field is 38888 [*karaṇīs*]. Half the product of these *karaṇīs* of the
area of the field and of the *karaṇīs* of the upward-side is the volume[202]. "Half",

---

[202]Another reading of this sentence could be: "Half the product of these *karaṇīs* which are the
area of the field and of the *karaṇīs* which are the upward-side is the volume".

Figure 10:

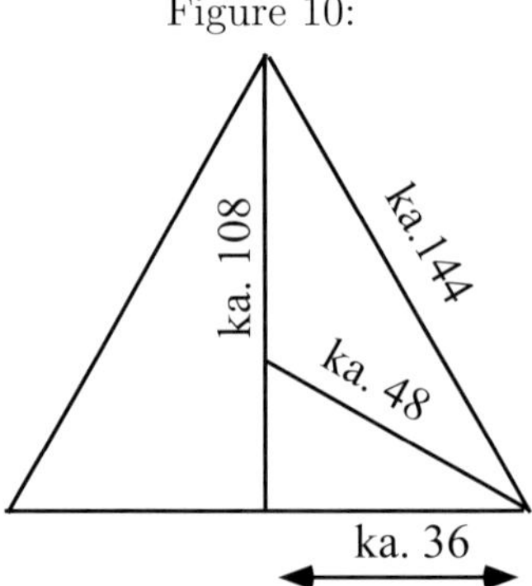

but here because the ⟨the product⟩ has the state of being a *karaṇī* ; ⟨the product⟩
is divided by the four *karaṇīs* of two. Because two ⟨should be⟩ *karaṇīs*, one should
divide by four *karaṇīs*. The volume obtained is 93312 *karaṇīs*.

An example:

2. Eighteen is the indicated value (*saṅkhyā*) of the ears of a *śṛṅgāṭaka*.|
Friend, I wish to know the remainder (i.e. the result) of the computation
of the upward ⟨of that *śṛṅgāṭaka* and its volume⟩||

Setting down :

Figure 11:

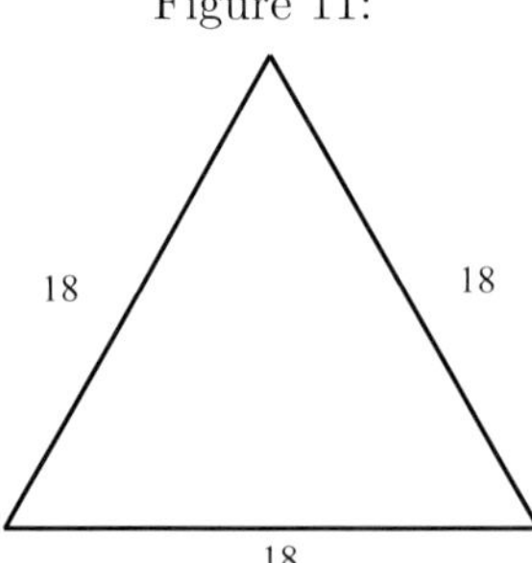

With the same procedure as before, the upward-side is 216 *karaṇīs*. The volume
too, obtained just as before, is 1062882 *karaṇīs*.

[Area of a circular field]

Now, in order to compute (*ānayana*) the area of a circular field (*vṛttakṣetra-phala*[203]), he states:

Ab.2.7.ab. Half of the even circumference multiplied by the semi-diameter, only, is the area of a circle|

     samapariṇāhasyārdhaṃ viṣkambhārdhahatam eva vṛttaphalam| 5

Pariṇāha is a circumference (*paridhi*). That which is and even (*sama*) and a circumference is an even circumference (*samapariṇāha* is a *karmadhāraya* compound). Its half.

Others, however, analyze the compound in another way: "That field which has an even circumference, that is *samapariṇāha* (a *bahuvrīhi* compound). Its half."

For those ⟨people⟩ the knowledge of half the area (*kṣetraphala*) follows, because the expression *samapariṇāha* ⟨being a *bahuvrīhi* and therefore⟩ referring to something else[204]would denote field[205].

Viṣkambha is a diameter (*vyāsa*). Half of the diameter is the semi-diameter (*viṣkambhārdha* is a genetive *tatpuruṣa*), multiplied by the semi-diameter is *viṣkambhārdhahatam* (an instrumental *tatpuruṣa*). It amounts to (paraphrasing *hata* by *guṇita*) multiplied by the semi-diameter.

One should accept that the syllable *eva* ("only") is used [206] in order to fill the   10 *āryā*. Or else, a restriction of the method (*upāya*) is made by using the *eva* syllable: "Half of the even circumference multiplied by the semi-diameter, only, is the area of the circle", that is, there is no other method.

---

[203]Another understanding of this compound would relate *kṣetra* to *phala* and we would thus read "In order to compute the area of a circle". The ambiguity rises firstly from the fact that an area can be called *kṣetraphala*, but is often abridged into *phala*. Secondly, *vṛtta* can be understood alternatively as an adjective meaning "circular", thus *vṛttakṣetra* is a "circular field"; or as a substantivated adjective, meaning "circle". Both readings appear in Bhāskara's commentary.

[204]Reading: *anyapadārthena* rather than *anyapādārthena* of the printed edition. This may be recalling *Āṣṣdtādhyāyi* 2.2.24:

    anekam anyapadārthe
    When ⟨a *bahuvrīhi*⟩ is referring to something else ⟨than its own constituents,
    then it can have⟩ more than one ⟨constituent⟩.

[205]In other words *samapariṇāha* is understood as a disk and not as the circumference. But we cannot imagine how the other half of the verse was interpreted.

[206]The expression used here is *karaṇa*, which means making, producing. The idea seems to be that a word is *made* into a verse and not used in it, maybe it also suggests that it is pronounced.

⟨Objection⟩

This is not so, because another method is heard of elsewhere: "The square of the semi-diameter with three as multiplier is the computation."[207]

This particular method is not accurate (*sūkṣma*), but practical (*vyāvahārika*). Therefore, there is only one method. There is no other for a computation in accurate mathematics (*sūkṣmagaṇita*).

15

An example:

> **1. I see accurately diameters (*viṣkambha*) ⟨whose lengths⟩ are eight, twelve and six. |**
> **Tell me, separately, the circumference (*paridhi*) and the area (*phala*) of those evenly circular ⟨fields⟩ (*samavṛtta*)‖**

Setting down: 8, 12, 6;

Figure 12:

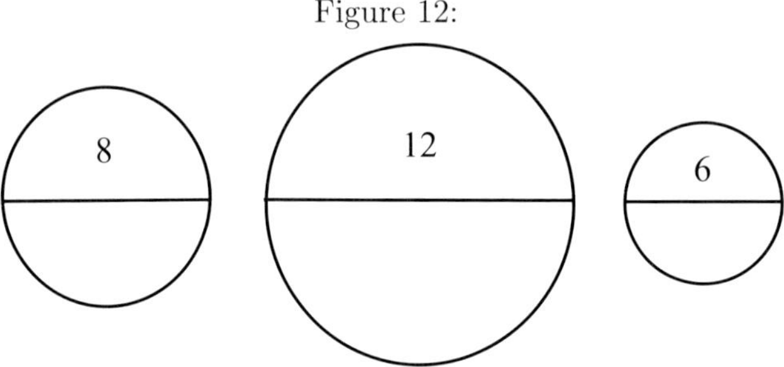

p.61, line 1   The circumferences obtained for these ⟨diameters⟩ by means of a Rule of Three, ⟨which uses⟩ as measure and fruit ⟨quantities⟩ (*pramāṇaphala*) the diameter and the circumference to be told [in Ab.2.10], are[208], in due order,

$$
\begin{array}{ccc}
25 & 37 & 18 \\
83 & 437 & 531 \\
625 & 625 & 625
\end{array}
$$

---

[207] *Vyāsārdhakṛtis trisaṅguṇā gaṇitam.*

[208] Knowing that a circle of diameter 20 000 has a circumference of 62832, we then have:

$$\mathcal{C}_1 = \frac{8 \times 62832}{20000} = \frac{15708}{625}$$

$$\mathcal{C}_2 = \frac{12 \times 62832}{20000} = \frac{23562}{625}$$

$$\mathcal{C}_3 = \frac{6 \times 62832}{20000} = \frac{11781}{625}$$

Procedure when computing the area: "half of the even circumference", the semi-diameter produced is 4.

Half of the even circumference of that ⟨circle⟩, which is $12\frac{354}{625}$, is multiplied by this very ⟨semi-diameter⟩.

The area of the circle produced is $166\frac{50}{625}$.

With just that procedure, the areas of the two remaining circumferences, are, in due order, $61\frac{113}{625}$ $343\frac{28}{1250}$.

[Volume of a circular solid]

p.61, line 15

In order to expose the volume (*ghanaphala*), he states:

Ab.2.7.cd. That multiplied by its own root is the volume of the circular solid without remainder.‖

tan nijamūlena hataṃ ghanagolaphalaṃ niravaśeṣam‖

This ⟨expression⟩ **that** expresses that area of a circular field which originates from the computation of the former half ⟨verse⟩. *Nijamūla* is one's own root. It amounts to: that which is the area of the field is multiplied by its own root. Or else, "that" is the area of the field, *nija* is "truly, not contradicted by tradition"[209]. ⟨As for:⟩ **multiplied by the root**: because another ⟨quantity⟩ has not been heard of, that area (*kṣetraphala*) is multiplied (*guṇita* paraphrasing *hata*) by its own root. Multiplied by its own root is *nijamūlahatam* (an instrumental *tatpuruṣa*)[210].

On the other hand, that area becomes a *karaṇī* when being made into a root (*mūlakrīyamāna*), because a root is [required] of a square (*karaṇī*). However, also, as there is no product of a *karaṇī* by a non-*karaṇī*, the area of the field is made into a *karaṇī*/the area of the field is squared (*karaṇyate*). Consequently, the following meaning is understood in fact: the square (*varga*) of the area of the field is multiplied by the area of the field.

---

[209]This explanation presupposes the reading *tan nijaṃ mūlena hataṃ* or *tan nijamūlahataṃ*, where *nijamūlahata* would be a *bahuvrīhi* beginning with an adverb. In both cases the reading is incompatible with the meter.

[210]Bhāskara does not here seem to be commenting on the expression actually used in the verse in Shukla's edition "*nijamūlena hatam*" but on the compound "*nijamūlahatam*" which is incompatible with the meter.

That which is and a solid and a *gola* (sphere) is *ghanagola* (a *karmadhāraya*); *gola* is a circular (*vṛtta*) ⟨object⟩[211]; **the volume of a circular solid** is *ghanagolaphala* (a genetive *tatpuruṣa*).

⟨As for:⟩ "**Without remainder**". Nothing remains with this computation (*karman*). With another computation, whith which they compute the volume of a circular solid, the volume of a circular solid is not produced without remainder, because that computation is practical (*vyāvahārika*):

> "When one has halved the cube of half the diameter multiplied by nine, the computation of the volume (*ghanagaṇita*) of the sphere (*ayoguḍa* lit. iron ball) ⟨is obtained⟩ |"

p.62, line 1
An example :

> **1. The diameters of ⟨three⟩ circles should be known as two, five, and ten in due order. |**
> **I wish to know the volumes of these circular solids succinctly‖**

Setting down:

Figure 13:

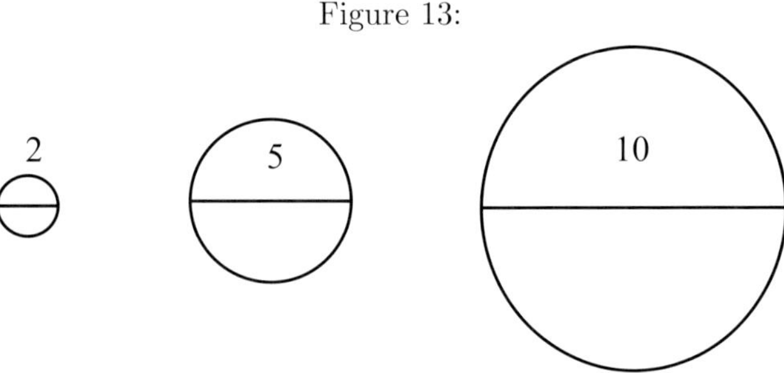

Their circumferences obtained with just a Rule of Three are, in due order,

$$\frac{6}{177}\frac{}{} \quad \frac{15}{177} \quad \frac{31}{52}$$

6	15	31
177	177	52	.
625	250	125

Procedure : With the previously told mathematical computation (*gaṇitakarman*) the area (*phala*) of the field [having two for] diameter is reached, $\dfrac{3}{177}$.
$\phantom{\dfrac{3}{177}}$1250

---

[211] The problem here is in Bhāskara's understanding of *gola/vṛtta* as an adjective meaning circular or as a noun meaning circle; in both cases the translation by "sphere" would seem absurd. I will adopt the following translation: *ghanagola* is "a circular solid", *gola* alone is "a sphere".

This very ⟨quantity⟩, which has become a *karaṇī* (*karaṇīgata*), should be considered to be its ⟨square-⟩root, because it is not a pure square (*aśuddhakṛti*). And that is made into the same category (*savarṇita*); what results is $\frac{3927}{1250}$.

This is multiplied by the square of the area; what results is the volume 31 *karaṇīs*, and $\frac{12683983}{1953125000}$ parts of a *karaṇī*.

Likewise, for the two remaining ⟨circles⟩ also, in due order the *karaṇīs* of the volumes and the parts of a *karaṇī* are: $\frac{7569}{7558983} \quad \frac{484476}{58983}$.
$$\frac{}{8000000} \quad \frac{}{125000}$$

15

[The area of a quadrilateral with equal perpendiculars]

p.63, line 1

In order to know the area and the size (*pramāṇa*) of the ⟨lines whose top is⟩ the intersection (*sampāta*)[212] – for [isosceles and scalene] quadrilaterals[213] and so on, and for interior ears (*antaḥkarṇa*), he states here an *āryā*:

> **Ab.2.8. The two sides, multiplied by the height ⟨and⟩ divided by their sum are the "two lines on their own fallings"[214].|**
> **When the height is multiplied by half the sum of both widths, one will know the area.‖**

5

> āyāmaguṇe pārśve tadyogahṛte svapātalekhe te|
> vistarayogārdhaguṇe jñeyaṃ kṣetraphalam āyāme‖

Āyāma, vistāra, dairghya are synonyms. Those two which are multiplied by the height are these *āyāmaguṇe* (a *bahuvrīhi* in the dual case). What are those? The sides (*pārśva*). One side is the earth (*bhū*), the other the face (*mukha*). The meaning is: the earth and the face (*vadana*) multiplied by the height.

The sum of these two is *tadyoga* (a genetive *tatpuruṣa*). ⟨The sum⟩ of which two? of the two sides. ⟨As for⟩ **"divided by their sum"** (*tadyogahṛte*). What are the two ⟨quantities divided⟩? The two sides having the height for multiplier (*ghna*).

A falling of one's own is *svapāta* (a genetive *tatpuruṣa*). Two lines on their own falling is *svapātalekhe* (a locative *tatpuruṣa* in the dual case). "Both are obtained separately" is the remaining ⟨part⟩ of the sentence. *Svapātalekhā* (the line on its own falling) is the name of the inner space (*antarāla*) ⟨delimited by⟩ the intersec-

10

---

[212]Please refer to the supplement for this verse (Volume II, section E) and to the Glossary (Volume II, section 5) for an understanding of, respectively, the segment this represents and a justification of this translation.

[213]The quadrilateral referred to here is a trapezium.

[214]Please refer to the supplement for this verse and to the Glossary at *svapātalekha*, for an understanding of, respectively, the segment this represents and a justification of this translation.

tion (*saṃpāta*) of the two interior ears and the middle of ⟨respectively⟩ the earth and the face.

Vistara is a width (*pṛthutva*) of the field.

⟨Objection⟩

If it was so, ⟨the word⟩ would be *"vistāra"* when a *ghañ* affix has been made ⟨by Pāṇini's rule⟩:

> When ⟨the root *stṛ*⟩ with the meaning of "spreading" is prefixed by *vi* and is not ⟨considered as⟩ a word ⟨the affix *ghañ* is added⟩[215]⟩ |'
> *Aṣṭādhyāyī; III. 3. 33*

This is not wrong. This ⟨word *stara*⟩, which is the end of the compound, with the word *vi* ⟨forms the compound⟩ *vistara* which is an abbreviation of *vivdhāstara*[216].

The sum of both widths is *vistarayoga* (a genetive *tatpuruṣa*), the meaning is: the sum of the earth and the face. Half the sum of both widths is *vistarayogārdha* (a genetive *tatpuruṣa*). That which is multiplied by half the sum of both widths is this *vistarayogārdhaguṇa* (a *bahuvrīhi*). What is ⟨multiplied by half the sum of both widths⟩? The height (*āyāma*). When that height is multiplied by half the sum of both widths, one will know the area (the second half verse is in the locative case). It amounts to: the height multiplied by half the sum of the widths is the area (*kṣetraphala*).

The size of the "lines on their own fallings" should be explained with the computation (*gaṇita*) of a Rule of Three on a field drawn by ⟨a person⟩ properly instructed [217]. Then, by means of just a Rule of Three with regard to the two sides which are a pair, the computation (*ānayana*) of ⟨the lines whose top is⟩ the intersection of the diagonals (*karṇa*) and a perpendicular (*avalambaka*) ⟨is performed⟩.

Here, with a previous rule (Ab.2.6.ab) the area of isoceles and uneven trilaterals should be shown. Or, with that rule which will be stated (Ab.2.9.) the computation of the area of the inner rectangular field ⟨should be performed⟩; in other fields as

---

[215] This rule accounts for the phonological making of the word *vistāra*.

[216] This sentence is so corrupt in the manuscripts that it is impossible to make any sense of it. The printed edition reads:

> *ayaṃ avastre staraśabdaḥ, tena viśabdena samāsānto "sau vividhastaro vistaraḥ*

    The first part of this sentence, is a reading proposed by the editor of the corrupt readings of the manuscripts. It would mean: "This word *'stara* in the meaning of *avastra*". However, the word *avastra*, "naked", has no meaning here. The manuscript readings given by the editor seem as senseless. Furthermore, the compound *samāsānta* that we have translated litteraly as "end of the compound" is usually understood as a technical term for compound suffixes. We have just rendered a general understanding here.

[217] Reading *samyagādiṣṭena* instead of the *samyaganādiṣṭena* (by ⟨a person⟩ improperly instructed) of the printed edition.

well, the acquisition of the ears, perpendiculars, etc., of those fields which are inside these other fields, ⟨should be made⟩ with just the characterisation that he (Āryabhaṭa) has taught. And no other procedure (*karaṇa*) will be ⟨required⟩ for these ⟨fields⟩ on account of the fact that they are situated in other ⟨fields⟩[218].

An example:

> **1. Let the earth (*bhūmi*) be fourteen, and the face (*vadana*) four units (*rūpa*)|**
> **The two chief (*agrau*) ears ⟨should measure⟩ thirteen, tell ⟨the lines⟩ whose top is the intersection (*sampātāgra*[219]) and the area (*phala*)||**

Setting down:

p.64,
line 1

Figure 14:

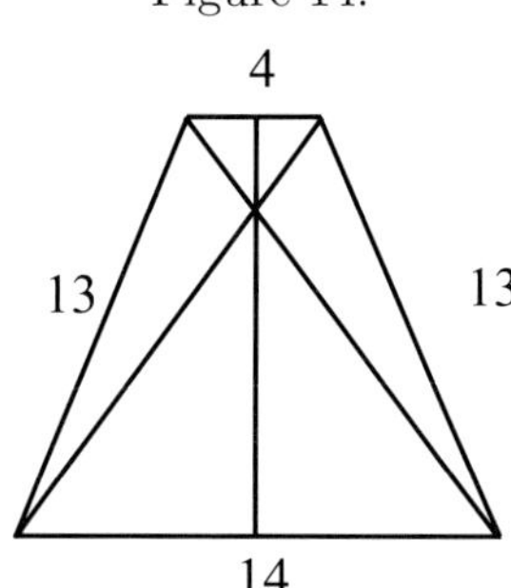

Procedure: The base ⟨of the inner right-angled trilaterals⟩ (*bhujā*) is half the difference of the earth and the face (*mukha*), [5]. A perpendicular (*avalambaka*) is established with that base, precisely by means of the computation (*gaṇita*) told separately (Ab.2.17 stated in BAB.2.6), and that ⟨perpendicular⟩ is 12. This very perpendicular is the height. Separately, the two sides are multiplied by it, what results is 48, 168. The sum of the sides is 18. The two quotients of the division by this ⟨last quantity⟩ are "the two lines on their own falling": $2\frac{2}{3}$ $1\frac{9}{3}$, half the sum of the widths is 9; the height multiplied by this is the area of the field, 108.

---

[218]The two paragraphs here may be suggestions of proofs/verifications of the procedures given by Āryabhaṭa in Ab.2.8. Concerning the areas, he explains that it is not because trilaterals, rectangles, etc., are inside another field (for example, a trapezium) that the rules given by Āryabhaṭa cannot apply. We have further developed this idea in the supplement for BAB.2.8.

[219]Even though the compound *sampātāgram* (top is the intersection) is in the singular case, we understand it as referring to the two segments (*lekha*, lines) which both have the same top at the intersection of the diagonals. We furthermore understand the adjective as modifying *lekha* as a masculine noun, and not as a femine, *lekhā*.

An example:

> **2. The numbers twenty increased by one, ten (*paṅkti*) and nine are mentioned|**
>
> For ⟨respectively⟩ **the earth (*dhātrī*), the ears and the face. Tell the computation ⟨of the field⟩ (*gaṇita*; i.e. the area) and "the lines on their own falling".‖**

10   Setting down :

Figure 15:

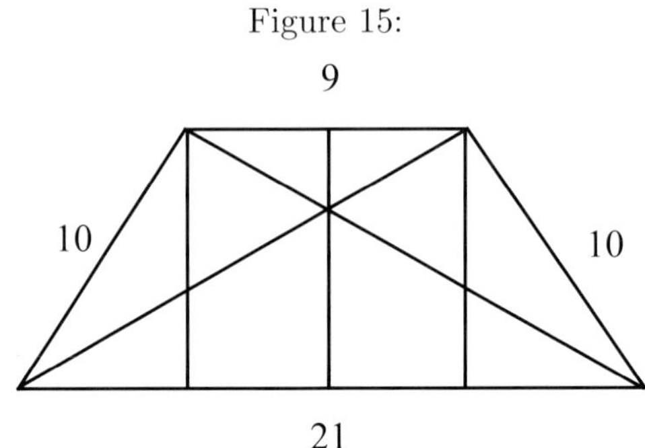

With the previous procedure, "the two lines on their own falling" are $\begin{array}{cc} 2 & 5 \\ 2 & 3 \\ 5 & 5 \end{array}$.

The area (*kṣetraphala*) is 120.

15   An example :

> **3. Thirty increased by three is the earth, the others are mentioned as seventeen|**
>
> **To what will amount the computation ⟨of the field⟩ and what will be the "two lines on their own falling"?‖**

p.65, line 1   Setting down :

For this tri-equi-quadrilateral (*trisamacaturaśra*), the two "lines on their own falling" obtained are: $\begin{array}{cc} 5 & 9 \\ 1 & 9 \\ 10 & 10 \end{array}$. The area (*kṣetraphala*) is 375.

In uneven quadrilateral fields, only the measure of the area is indicated, not the two

5   "lines on their own falling", because the perpendicular (*avalambaka*) is difficult to know.

Another ⟨specification⟩ also: The uneven quadrilateral field here is not the same as the fields ⟨called "uneven"⟩ in other ⟨treatises⟩ on mathematics (*gaṇita*).

Figure 16:

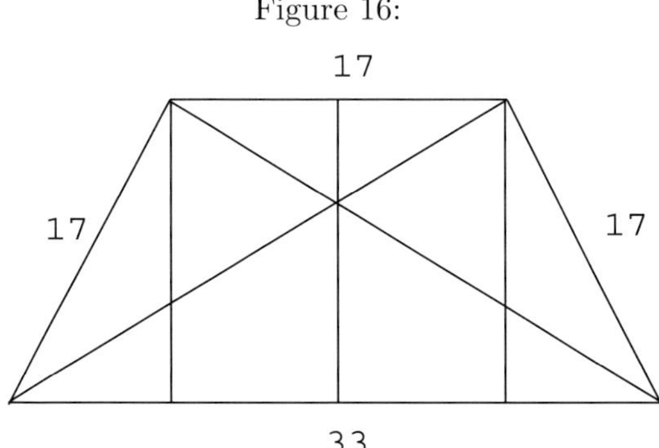

**4. It is told that the size of the earth (*vasudhā*) is sixty fitted together
with the face which is the square (*kṛti*) of five|
The two ears are measured by thirteen multiplied respectively by four
and three.‖**

The two perpendiculars of this ⟨quadrilateral⟩ are not equal (*na sadṛśa*). The
two perpendiculars of the ⟨field⟩ which is instructed here (in Ab.2.8) have the
same value (*tulyasaṅkhya*). Therefore, there is a difference between this ⟨uneven-
quadrilateral treated in Ab.2.8⟩ and the uneven-quadrilateral field instructed in
other treatises on mathematics, even though unevenness exists ⟨in both cases⟩[220].

Now ⟨concerning⟩ that uneven-quadrilateral-field taught in a different treatise on
mathematics and the one which is taught here (i.e quadrilateral fields which have
equal perpendiculars), the specification of the area of these very two ⟨types of
fields⟩ can be [made] with this teaching (i.e. the one given in Ab.2.8.cd) as well.

What is ⟨taught in the case⟩ of a ⟨field⟩ whose perpendicular (*avalambaka*) is
difficult to know?

It is replied: "In uneven fields, only the measure of the area [is] indicated and not   15
the two 'lines on their own falling'." Now if the perpendicular is known, then one
can know and the area and the "lines on their own falling". How? With the very
mathematical procedure explained previously.

---

[220]In other words, the trapezium is defined by Bhāskara as a quadrilateral with equal perpendiculars; the "uneven-quadrilatera" of this treatise is a non-isoceles trapezium. However in other
treatises, an "uneven quadrilateral", which does not have equal perpendiculars is any quadrilateral. More precisely, Bhāskara considers that the two equal perpendiculars should always be
inside the field...

p.66,   An example :
line 1

    5. The height is told to be twelve, the earth is twenty minus one|
    The face is told to be five; now the two ears of that ⟨field⟩ are said to
        be ‖
    Ten increased respectively by five and three.|
    I wish to know accurately the area and the two "lines on their own
5   **falling".‖**

Setting down:

Figure 17:

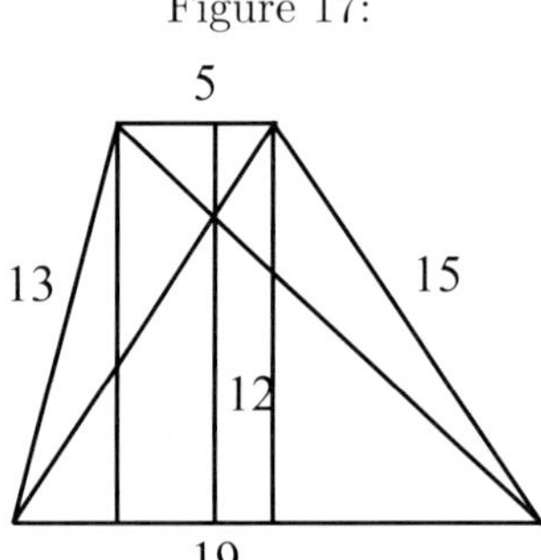

The two "lines on their own falling" obtained are $\begin{smallmatrix} 2 \\ 1 \\ 2 \end{smallmatrix}\ \begin{smallmatrix} 9 \\ 1 \\ 2 \end{smallmatrix}$. The area is 144.

The computation of the area (*phala*) and the computation of the "lines on their own falling" are just like this in other similar kinds of fields also.

p.66,   [Verification and area]
line 10

For all fields, aiming at the areas (*phala*) and ⟨their⟩ verification (*pratyayakaraṇa*), he states:

    Ab.2.9-ab. For all fields, when one has acquired the two sides, the area
        is their product |

    sarveṣāṃ kṣetrāṇāṃ prasādhya pārśve phalaṃ tadabhyāsaḥ|

For all fields, the area is to be indicated.

How?

"When one has acquired (*prasādhya*) the two sides".

The sound *"pra"* expresses "specified" (*prakṛṣṭa*), that is, when one has acquired the two sides with specificity (*prakarṣeṇa*)[221].

⟨Question⟩

Then what is the specificity of the two sides that are being acquired?

It is replied: sideness (*pārśvatā*[222]).

What, then, is the meaning of the word "sideness"?   15

It is replied: When ⟨the area of⟩ all fields is being acquired, ⟨then⟩ it is only on the side; it amounts to: ⟨acquiring⟩ a rectangle[223].

⟨As for:⟩ **"the area is their product"**. The area of all of these fields, ⟨which amount to⟩ the rectangles whose sides have been introduced, is the product of these two sides; it amounts to: the product of the width (*vistāra*) and the length (*āyāma*).

Abhyāsa, guṇanā and *saṃvarga* are synonyms[224].

⟨Objection⟩

Now, because the word "all" expresses "everything without exception" (*nira-*   p.67,
vaśeṣa), every field, indeed, without exception is referred to (*ākṣipyante*) ⟨with   line 1
this rule⟩. Therefore, because the area of all fields has been established with this
very rule, the mentioning of the previously stated rules (i.e., Ab.2.6.ab, Ab.2.7ab,
Ab.2.8) was useless.

It is not useless. The verification and the ⟨computation of⟩ the area are told with
this ⟨rule⟩.

---

[221] A problem arises here, as both the past passive participle *prakṛṣṭa* and the active noun *prakarṣa* derived from the root *prakṛṣ-*, convey generally the meaning "excellency". However, this would be meaningless in the discussion that follows. One of the particular meanings of *prakarṣa* is "specialty".

[222] K.V. Sarma has in a personal communication proposed to understand the property described by the word *pārśvatā* as "orthogonality". This is what would distinguish one side from another.

[223] This part is so elliptic that it remains obscure. Since *pārśva* cannot be a nominative because it is transformed by *saṃdhi* with the following *eva* (it should keep as a nominative the form *pārśvaḥ*, or in the dual *pārśve*) we read it as a locative singular. We could not make sense of the sentence, even when considering the editors addition. This is a literal translation, without any additions:

> *yadi sarvakṣetram prasādhyamānam, [tadā "pārvatā"-abdāsya artha] pārve*
> *eva bhavati, āyatacaturaram eva iti yāvat —*
> When all fields are being acquired, [then the meaning of the word "sideness"
> is] it is only on the side; it amounts to: a rectangle.

[224] This list was already given in BAB.2.3.ab.

Verifications of the areas of the ⟨previously⟩ stated fields ⟨are made⟩ since the experts in mathematics Maskari, Pūraṇa, Pūtana, etc., verify the area of all fields in[225] rectangular fields. And it has been said:

Always, when one has sought (*anugamya*) the area by means of told procedures, then, one should know ⟨its⟩ |
verification in a rectangular field, since in rectangles the area is obvious (*vyakta*)‖

The computation (*ānayana*) of the areas of fields unstated ⟨in known rules is possible⟩ just by transforming (*karaṇa*) the desired field into a rectangle.[226]

⟨Question⟩

But how are the computation of the area and the verification acquired with just one effort? Now, if this ⟨rule⟩ was originally made in order ⟨to carry out⟩ a verification, how can that ⟨rule⟩ be for the computation of the area ? And, ⟨conversely, if this rule was originally made⟩ in order to compute the area, how can ⟨it be used⟩ for a verification?

There is no drawback. It has been observed that what is originally made for one purpose is the instrument (*sādhaka*) of another purpose. That is as in the following ⟨case⟩:

Canals are constructed for the sake of rice paddies. And from these ⟨canals⟩ water is drunk and bathed in.
Aṣṭādhyāyī, 1. 1. 22, *Pātañjalabhāṣyam*.

It is just like this in this case also. It is as follows:

An example concerning the computation of the area of a rectangular field:

1. How much is the computation of those rectangles (i.e. their areas) whose widths are ⟨respectively ⟩ |
Eight, five and ten (*paṅkti*) and also whose lengths (*dairghya*) are sixteen (*aṣṭi*), twelve and fourteen (*manu*)?‖

Setting down:

Eight is one side, sixteen is the other. The product of both sides ⟨is made⟩, the area brought forth is 128. For the remaining two also, it is just like that ⟨the results are respectively⟩ 60, 140.

---

[225] The peculiar use of the locative here, may be referring to the fact that fields whose areas are known are transformed *into* rectangles. This is what a verification seems to consist in here.
[226] This paragraph is translated in [Hayashi 1995 ; p. 73 and p. 74] and in [Shukla 1976; p.liv.].

Figure 18:

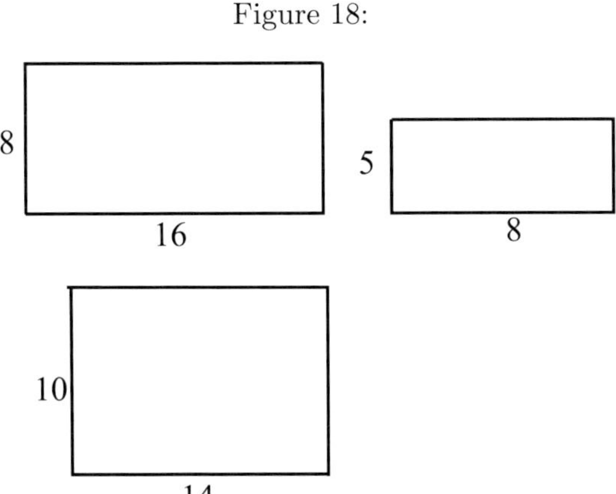

The   verification of the areas of the fields brought about with previous rules is   p.68,
shown. It is as follows: line 1

> **2. The areas of tri- and quadri-laterals, and of circles were seen with
>   computational (*gaṇita*) ⟨rules⟩ |**
> **Say how the verification of all of these ⟨fields⟩ is produced.‖**

How is ⟨produced⟩ the verification of the area of just that equi-trilateral field
previously seen (in example 1 of BAB.2.6.ab)?

Setting down :   5

Figure 19:

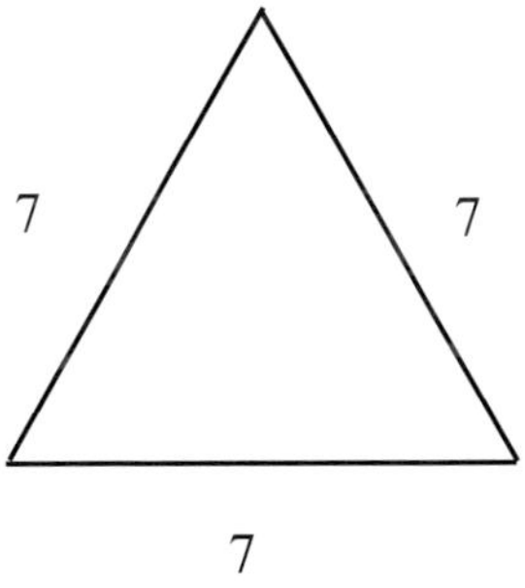

Just this has been scattered[227] ⟨and rearranged⟩, what has been produced is a
rectangular field:

---

[227]Reading *vyastam* as in all manuscripts rather than *nyastam* as in the printed edition.

Figure 20:

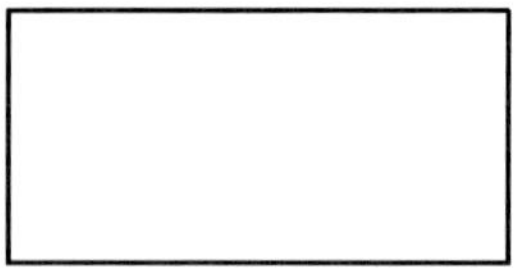

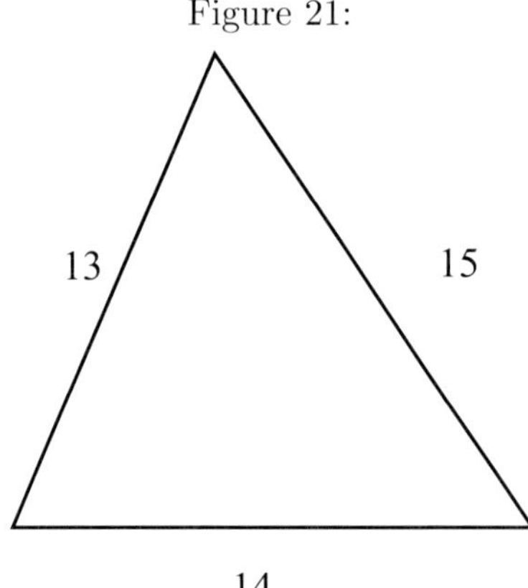

[The perpendicular of that trilateral, which is the length ⟨of the rectangle⟩] is $3\tfrac{36}{4}$ *karaṇīs.*

10   [Half the base   which is the width ⟨of the rectangle⟩] is $1\tfrac{12}{4}$ *karaṇīs.*

Since the area is the product of the two sides, it is $3\tfrac{450}{16}$ *karaṇīs* as previously written.

⟨The verification⟩ is just like that in isosceles and uneven ⟨trilaterals also⟩. Setting down the previously told uneven ⟨trilateral⟩[228]:

Figure 21:

Once again its perpendicular is the length ⟨of the rectangle⟩, 12. Half the base is the width, 7.

p.69, line 1

Just as before, for this ⟨field⟩ too the area is the product of the width and the length, 84.

---

[228]Figure 21 inverts the sides measuring 13 and 15 of figure 7 of BAB.2.6.ab. If this is the case in manuscripts, it would show that both are considered as the same triangle.

Figure 22:

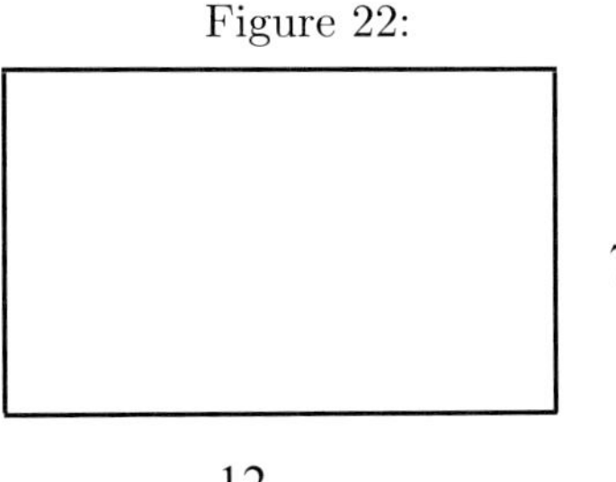

Or else, its area is the sum of half the areas (*kṣetraphala*) of two rectangular fields. This ⟨trilateral's⟩ area is the sum of half the areas of these two ⟨rectangles⟩, the one whose width is five and length twelve, and the second one also, whose width is nine and length twelve.

Setting down these two fields, the first one whose width is five and length twelve, and also the second whose width is nine and length twelve:

Figure 23:

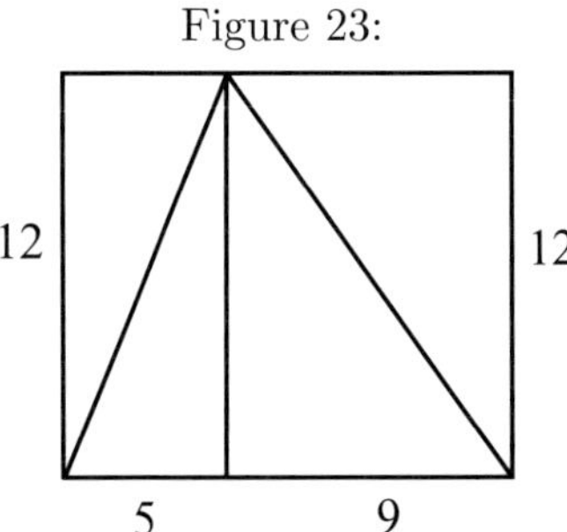

The area of the twelve by five ⟨field obtained⟩, with the method (*krama*) of the product of the width by the length, is 60; its half exactly ⟨which has been fitted in⟩ this uneven-trilateral field, 30; the area of the field whose width is nine and length twelve is 108, its half also has been fitted in this ⟨uneven trilateral⟩ 54. The sum of these two half areas, is once again the same ⟨as the area computed before⟩, eighty-four, 84.

Likewise in isosceles, tri-equi and uneven quadrilaterals also the area should be verified (*pratyāyanīya*).

In a circular field, the semi-diameter (*viṣkambhārdha*) is the width, half the circumference (*paridhi*) is the length, just that ⟨gives⟩ the rectangular field.

In this way, with one's own intelligence the area is inferred in miscellaneous fields. It is as follows:

3. The face is seen as eleven and then the opposite face is said to be nine|

The height (*āyāma*) is twenty. What should be its area, calculator (*gaṇaka*)?‖

15

Setting down:

Figure 24:

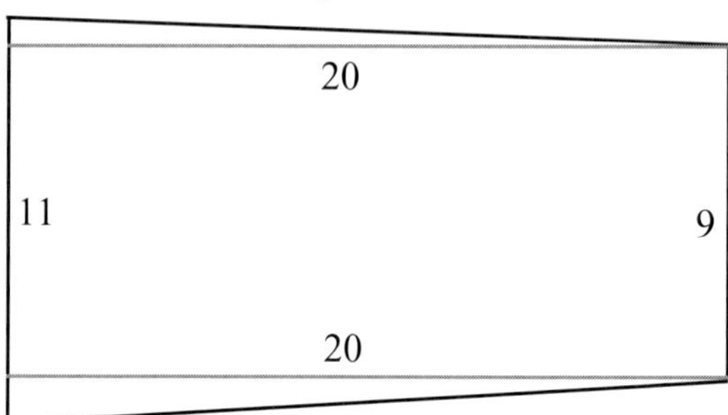

p.70, line 1   Procedure : "When one has acquired the two sides, the area is their product", the sum of the two unequal sides is 20, its half is 10. [The length ⟨of the rectangle⟩ is 20]. The sides ⟨of the rectangle⟩ are these two ⟨quantities⟩, ten and twenty. The area is their product [200].

An example:

4. The two faces of a drum (*paṇava*) are eight and eight, the separation (*vyāsa*) is two, and the length (*dairghya*) is said to be sixteen. |

5

One should say to what amounts the area (*phala*) of that ⟨field⟩ which is shaped in the form of a drum.‖

Setting down:

Procedure: The sum of the two faces is 16, its half is 8. This increased by the width (*vistāra*), 2, is 10. Its half is 5. "When one has acquired the two sides, the area is their product", in this way, the area has been brought forth, 80.

10

An example:

5. The width is said to be five, the belly (*udara*) is nine, its back (*pṛṣṭha*) fifteen|

To what amounts the area of the tusk-field (*karidantakṣetra*)? This should be indicated clearly.‖

Figure 25:

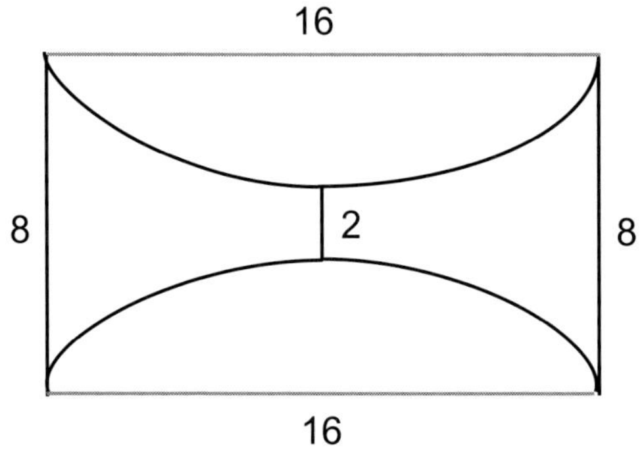

Setting down[229]:

Figure 26:

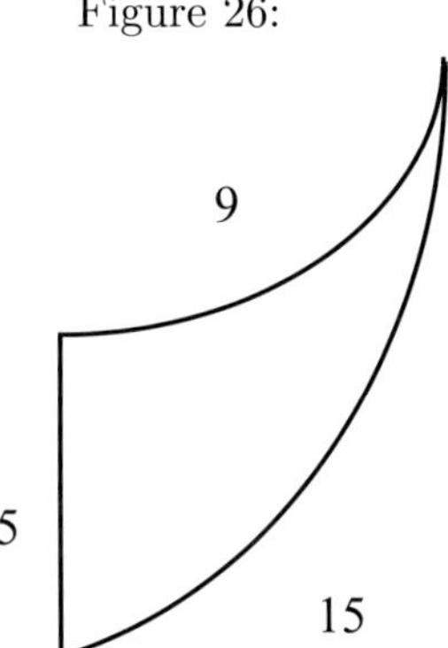

Procedure : The sum of the belly and the back is 24, the half 12. This multiplied by half the width is the area, which is thirty, 30. p.71, line 1

Likewise, in all fields, by ⟨properly⟩ choosing the two sides, the area should be indicated.

---

[229]The diagram of the edition erroneously has 14 indicating the length of the back.

[A chord equal to the semi-diameter]

In order to exhibit a chord (*jyā*) equal to the semi-diameter of an evenly-circular
⟨field⟩, he states :

**Ab.2.9.cd. The chord of a sixth part of the circumference, that is equal
to the semi-diameter‖**

paridheḥ ṣaḍbhāgajyā viṣkambhārdhena sā tulyā‖

Paridhi, pariṇāha and *vṛtta* are synonyms. That, which is the chord of a sixth
part (*ṣaḍbhāgajyā* is a genetive *tatpuruṣa*) of that circumference, is equal to the
semi-diameter.

The sixth part of the circumference is a pair of *rāśis* [230]. That chord which subtends
(*avagāhin*) a two-*rāśi* field is the chord of the sixth part of the circumference. Half
of that is the half-chord (*ardhajyā*) of one *rāśi*. And all this should be explained
in a diagram (*chedyaka*). And in this diagram, which is drawn with a compass
(*karkaṭa*) with a secured (*sita*) sharp stick (*vartyaṅkura*) fastened to the mouth-
spot (*mukhadeśa*), that which is half of the chord of a sixth part is the half-chord
of a *rāśi*. The production of unknown half-chords will be told (in Ab.2.11.) with
that half-chord [231].

In a circular field, six equi-trilateral fields have been shown incidentally (*prasaṅ-
gena*) by one who wishes to explain this very chord of the sixth part. In this case
the sides (*bāhu*) are the semi-diameter. Or there are six bow-fields (*dhanuḥkṣetra*)
whose chords (*jyā*) are the semi-diameter. And likewise, there is a hexa-edged field.
And the use of this exhibition of the chord of the sixth part will be told in this
verse (*kārika*): "One should divide the quarter of the circumference of an evenly
circular ⟨field⟩" (⟨Ab.2.11.ab⟩).

p.71,   [Relation of the diameter with the circumference in a circle]
line 17

In order to compute (*ānayana*) an evenly-circular ⟨field⟩ (*samavṛtta*) with a Rule
of Three, he states:

---

[230] Please see the supplement for this commentary (Volume II, section F on page 40), for an
explanation of this statement and the following. This sentence can also be translated as: "The
sixth part of the circumference is a two-*rā'si* ⟨field⟩."

[231] Reading *tayā "rdhajyāyānirjñātāyāḥ ardhajyāyāḥ utpattim* rather than the reading of the
printed edition or of the manuscripts.

**10. A hundred increased by four, multiplied by eight, and also sixty-two
thousand|**

Is an approximate circumference of a circle whose diameter is two *ayu-
tas* [232]**||**   20

caturadhikaṃ 'satam aṣṭaguṇam dvāṣaṣṭis tathā sahasrāṇām|
ayutadvayaviṣkambhasyāsanno vṛttapariṇāhaḥ||

Caturadhikam is **increased by four** (an instrumental *tatpuruṣa*). What is that
⟨which is increased by four⟩? **A hundred.** *Aṣṭaguṇa* is **multiplied by eight.** This is
what is told: "Eight hundred and thirty-two". And sixty-two thousand. Both are
summed (*ekatra*): 62832.

Ayutadvayaviṣkambha is ⟨that which is⟩ and a diameter and of two *ayutas* (a
karmadhāraya). Or else the diameter whose number (*saṅkhyā*) is two *ayutas* or
whose size (*pramāṇa*) is two *ayutas* is a "**diameter of two** *ayutas*". Whose diameter
is two *ayutas*. And that is 20000.   p.72,

                                                                                  line 1
"Approximate" (*āsanna*) is near.

⟨Question⟩

What is it an approximation (*āsanna*) of?

Of the accurate (*sūkṣma*) circumference.

⟨Question⟩

How is it known as an approximation of an accurate ⟨value⟩ (*sūkṣmasya āsanna*)
and not indeed as an approximation of a practical ⟨value⟩ (*vyāvahārikasya āsanna*),
as long as the determination of what has been heard ⟨in the verse⟩ is the same
(*tulya*) ⟨whether the value approximated is⟩ accurate or practical (i.e. in all cases
the value is an approximation).

There is no mistake. This is just a doubt (*sandeha*). And[233] the ⟨following⟩ stands
for all doubts: "A specific meaning arises from an ⟨authoritative⟩ explanation
(*vyākhyāna*), [by no means does ⟨a rule⟩ become invalid (*alakṣaṇa*) because of a
doubt]". (*Aṣṭādhyāyī, Śivasūtra 6, Pātañjalabhāṣyam*)   5

Therefore we are giving the ⟨authoritative⟩ explanation that it is the approximation of an accurate ⟨value⟩.

Or else, the word *āsanna*, which means "near that", denotes. Therefore, that
⟨accurate value⟩ itself is ⟨also⟩ expressed with the word *āsanna*. It is then, somewhat different (*bhinna*) ⟨from the approximate value⟩.[234] If it were an approxima-

---

[232] As stated in AB.2.2., an *ayuta* is the name of ten thousand.

[233] Reading *sarvasandeheṣu cedam* instead of the *sarvasandeheṣu vedam* of the printed edition.

[234] This last sentence is somewhat difficult to understand. Its meaning may be that there is a
fractional part separating the approximate value from the accurate one. Indeed, *bhinna* is the
word usually used to denote an integer increased or decreased by a fraction. Also the demonstrative used to refer to what we suppose is the "accurate value" is in the neuter, and not in the
feminine or the masculine as it should be if it was refering to *sūkṣmasaṅkhyā* or *sūkṣmaparidhiḥ*.

tion of a practical ⟨value⟩, then the circumference ⟨obtained⟩ from that practical ⟨value⟩ would be even worse. No one would make an effort ⟨leading to something⟩ worse. Therefore it has been established by ⟨this very⟩ rule ⟨of common sense⟩ (*nyāyasiddham*) that it is an approximation of an accurate ⟨value⟩.

⟨Question⟩

Now, why is the approximate circumference told, and not indeed the true circumference (*sphuṭaparidhi*) itself ?

They[235] believe the following: There is no such method (*upāya*) by which the accurate circumference is computed.

⟨Objection⟩

But here it is:

The *karaṇī* which is ten times the square of the diameter is the circumference of the circle|[236]

In this case also, this ⟨rule⟩ "the circumference of a unity-diameter ⟨circle⟩ is ten *karaṇī̄s*[237]' is merely a tradition (*āgama*) and not a proof (*upapatti*) .

⟨Objection⟩

Now, some think that the circumference of a field with a unity-diameter when measured directly (*pratyakṣa*) is ten *karaṇīs*.

This is not so because *karaṇīs* do not have a statable size.

⟨Objection⟩

The circumference ⟨of the field⟩ with that ⟨unity-⟩diameter, when enclosed by the diagonal, whose *karaṇīs* are precisely ten, of a rectangular field whose width (*vistāra*) and length (*āyāta*) are respectively one and three, that ⟨circumference⟩ has that size (i.e. it measures ten *karaṇīs*).

But that also should be established (*sādhya*).

And something else : In a circular field there are four bow-fields and one rectangular field. The sum (*samāsa*) of their areas must be the area of the circular field. These areas ⟨when computed with the ten *karaṇīs*, and⟩ summed (*saṃyojyamāna*) do not equal the area of the circular field.

---

[235] This anonymous collective voice is used from time to time in this commentary, as in BAB.2.3, and must be referring to scholars who had commented on this point.

[236] This verse is given in *prākṛta* in the edition; a Sanskrit translation, supplied by the editor appears within brackets under it. For a discussion on this verse and the following, please see the supplement for verse 10.

[237] This is expressed by a plural, *da´sakaranyaḥ*. It may be understood as if the *karaṇī* was a sort of measure of square numbers, thus being in the plural form. Throughout this part of the commentary, this is the form used.

An example in order to explain (*pratipādana*) that:

> **In a field whose diameter is ten, in the eastern (*pūrva*) and western (*apara*) part, having penetrated (*avagāhya*) one unit (*rūpa*)|**
> **The chord (*jīvā*) is six. And in the north (*uttara*) and south (*dakṣina*) having penetrated two units ⟨the chord is⟩ eight.‖**

A *gāthā*, which is a rule ⟨giving⟩ a method to compute these chords (*jīvā*) :

p.73, line 1

> **The diameter decreased by the penetration (*avagāha*) should be multiplied by the penetration|**
> **Then the root of the product multiplied by four is the chord of all fields‖** [238]

5

And setting down the bow-fields[239]:

Figure 27:

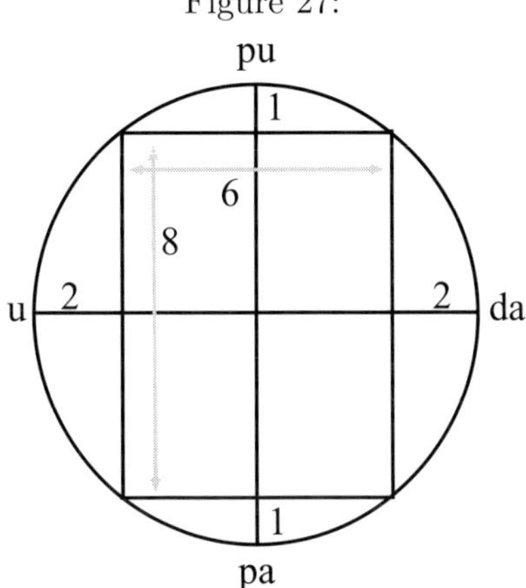

A *gāthā* which is a rule for computing the area of a bow-field:

10

> **Having multiplied by ten *karaṇīs*[240] the chord multiplied by the quarter of the penetrationon, ⟨this⟩ will be the area|**
> **In that field which is a strip like a bow, this procedure (*karaṇa*) should be known‖**

---

[238] Once again the verse is in *prākṛta* and translated within brackets in Sanskrit.

[239] Concerning cardinal directions, one can refer to Volume II, Supplement to verse 11 H.1.2.

[240] This verse is in *prākṛta* and translated within brackets in Sanskrit. The *prākṛta* reads here *dasikaraṇi*, which seems to be a singular form, Shukla translates with *daśakaraṇībhir*, in the plural form.

The two areas of the eastern and western bow-fields are indeed, according to this *gāthā* ka. $\dfrac{90}{4}$, ka. $\dfrac{90}{4}$.[241] These two areas should be added using a method (*vidhāna*) to sum (*prakṣepa*) *karaṇīs*[242].

A *gāthā* which is a rule for the sum of *karaṇīs*:

**When one has reduced (*apavartya/auvaṭṭi*) ⟨the two *karaṇīs* to be summed⟩
by ten, then, the sum (*samāsa*) of the roots ⟨of the results is taken⟩.
That which arises from the same (i.e., the square of the) ⟨sum⟩ is|
Multiplied by the digits of the reducer (i.e. ten), ⟨the result is a *karaṇī*;
in this way⟩ the sum of ⟨two⟩ *karaṇīs* should be known.‖**

Once it is done, the result is ka. 90. And then for the southern and northern bows, in accurately the same way, the areas (*phala*) are ka. 160, ka. 160. [And the sum (*samāsa*)] is ka. 640. The sum of the two sums (*samasta*) is indeed ka. 1210. The area of the rectangular field standing in the middle is 2304 *karaṇīs*.

When summing, with the method (*kriyā*) to sum *karaṇīs*, the quantity which is the sum of the areas of the bow fields and this (i.e., the area of the rectangle), both quantities are unsummable(*akṣepatā*).[243]

And the computation of the back ⟨of a bow field⟩ (*pṛṣṭha* i.e. the arc) also, ⟨when⟩ used for a determination of the calculation (*prakriyāparikalpanā*) of a circumference with ten *karaṇīs*, does not always [work, because] when computing the back ⟨of a bow field⟩, ⟨the following⟩ half *āryā* rule ⟨is used⟩:

**The sum of a half arrow and the quarter of ⟨its⟩ chord, multiplied to
itself, [with ten as a multiplier, these are the *karaṇīs* ⟨that measure
the back of the bow field⟩]|**

[Here is an example:

**When one has penetrated by two ⟨units in a circle⟩ whose diameter is
fifty-two, ⟨what is the chord and corresponding arc?⟩ |]**

According to that ⟨rule⟩:"the diameter decreased by the penetration", the chord obtained is twenty [20]. [With this chord], the computation of the arc ⟨is undertaken⟩: the quarter-chord is [$\dfrac{20}{4}$ =] 5; half the arrow (*śara*) is [1], the sum is 6.

Multiplied to itself: 36. Multiplied by ten: 360. These *karaṇīs* are the back ⟨of the bow-field⟩.

The square of the whole chord is four hundred and the back is three hundred and sixty for *karaṇīs* : how is this possible? The back ⟨of the bow-field⟩ must be bigger

p.74, line 1
5
10
p.75, line 1

---

[241]The "ka." in the printed edition is to be understood as an abbreviation for *karaṇī*.

[242]A short "i", as it appears in the printed edition, is wrong here–probably a misprint.

[243]In this case ka. 2304 cannot be reduced with 10 into an integer while using the method for summing *karaṇīs*, so the rule cannot be applied.

than the chord.

Therefore, this back ⟨of the bow-field⟩, which is reflected upon by by those who follow exceedingly accurate (*sūkṣma*) ⟨procedures⟩, happens to be shorter than the chord. Therefore, we pay homage to the ten *karaṇīs*, which seem attractive ⟨but are⟩ misconceived.

Now, another example again :

In a field whose diameter is twenty-six, having penetrated one; ⟨how much is the chord and the back of the bow-field?⟩.

With accurately the same procedure as before, the chord is ten, 10. Just as previously, the back ⟨of the bow-field⟩ is ninety for *karaṇīs* (*navatiḥ karaṇīnām*), 90. The square of the chord is one hundred, 100.

When seen in this way this ⟨value⟩ appears as being exceedingly rough (*atyantasthūla*). Therefore it has been pointed out correctly that there is no such method ⟨by means of which the accurate circumference is computed⟩.

⟨Objection⟩

Now, why are these two big quantities (20000 and 62832) told and not their reduced form[244] (1250 and 3927 for instance). The teacher is fond of brevity (*laghavika*), for such a lover of brevity it is not appropriate to state such large quantities.

Excuse in this only case the teacher.

Or else "a diameter of two *ayutas*" is told with a small amount of letters. But when one states a reduced diameter it does not have the state of having a small amount of letters.

Or else ⟨some⟩ think: "When stating a large circumference and diameter, when the chords of large diameters have small increases or losses there is no difference in the result because the difference is small". And thus in ⟨the table of sinus differences⟩ beginning with *makhi* (225[245]), sometimes incorrect ⟨values⟩ (*asata*) are adopted, sometimes correct ones (*sata*) are rejected[246].

---

[244]Reading *apavartitāu* instead of *apartitāu* of the printed edition.

[245]This way of expressing numbers is defined by Āryabhaṭa in Ab.1.2 and subsequently used when producing versified tables, such as the one referred to here which is given in Ab.1.12. As this way of naming numbers is not used in the *gaṇitapāda* we have not described them here. There has been a relatively abundant literature on the subject. For a final word, one can refer to [Shukla & Sharma 1976; p. 3-5]. [Wish 1827; pp. 55-56] is the first to point out it's existence. For controversies one can look up [Kaye 1908; p. 116-119], [Fleet 1912; p. 459-461], [Ganguly 1927] [Chaṭṭopādhyāya 1927; pp. 110-115], [Sen 1963; 298-302].

[246]In his commentary to the two following verses, Bhāskara attempts to reconstruct the table of half-chords given by Āryabhaṭa in the verse 12 of the first part of the *Āryabhaṭīya* (Ab.1.12). From the way he proceeds, that is described in the Supplements for BAB.2.11 and BAB.2.12, evidently approximations are used. Takao Hayashi in [Hayashi 1997] has also examined closely the different approximations used by Bhāskara and others, and also by Āryabhaṭa himself in the establishment of this table of Rsinuses.

Pariṇāha is a circumference (*paridhi*). *Vṛtta* (circle) is a field. *Vṛttapariṇāha* is the circumference of a circle (a genetive tatpuruṣa). The meaning is: the circumference (*paridhi*) of a circle.

With this ⟨rule⟩ when the diameter is known, the circumference (*paridhi*) is computed and when the circumference is known, the diameter ⟨is computed⟩.

How?

If this is the circumference of that diameter, how much is ⟨the circumference⟩ of the diameter which is the desire (*icchā*)? And if that has this circumference, how much is the ⟨diameter⟩ of that circumference which is the desire? [247]

An example:

> **1. Compute, friend, in accordance with *gaṇita*, the computation of ⟨the circumference⟩ of circles|**
> **which are near the accurate ones, whose diameters, are respectively two, four, seven and eight.‖**

Figure 28

Setting down the field:

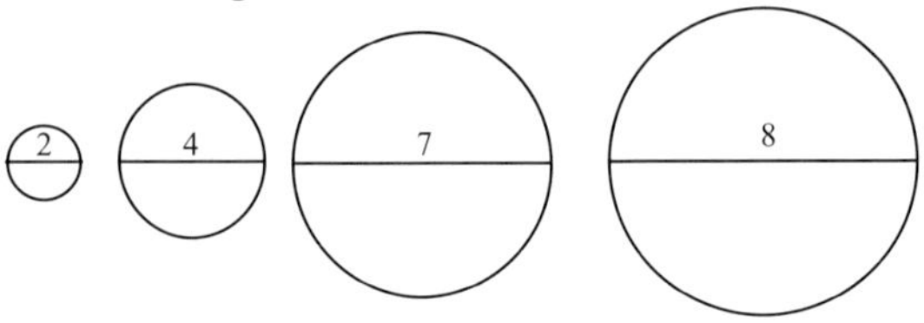

The circles obtained are, in due order:

6	12	21	25
177	354	1239	83
625	625	1250	625

An example when computing the diameter knowing the circumference:

> **2. Compute and tell me the diameters ⟨of these circles whose circumference are⟩ nine-nine-two (*yama*)-three (*Rāma*) decreased by eight parts of five (*śara*)-two (*yama*)|**
> **And zero (*kha*)-zero-six (*rasa*)-twenty-one (*vṛnda*)[248].‖**

Setting down:

$$\begin{array}{c} 3299 \\ 8° \\ 25 \end{array} \Big| \; 21600:$$

The two diameters (*vyāsa*) obtained are in due order:

$$\begin{array}{c|c} 1050 & 6875 \\ & 625 \\ & 1309 \end{array}$$

---

[247] A way of stating a Rule of Three.

[248] According to the setting down this should be a name for 21. It actually occurs in the *Āryabhaṭīya*, in Ab.2.2 as 10^9.

p.77,
line 1

[Computation of ⟨half-⟩chords using an operation in a diagram]

Now, in order to compute chords, he states:

> **Ab.2.11. One should divide the quarter of the circumference of an**
> **evenly-circular ⟨field⟩. And, from trilaterals and quadrilaterals|**
> **As many half-chords of an even ⟨number of⟩ unit arcs as one desires**
> **⟨are produced⟩, on the semi-diameter.‖**[249]
> **samavṛttaparidhipādaṃ chindyāt tribhujāc caturbhujāc caiva|**
> **samacāpajyārdhāni tu viṣkambhārdhe yatheṣṭāni‖**

Samavṛttaparidhi is that field which has an evenly circular circumference (a *bahu-vrīhi* to which the word *kṣetra* (field) may be supplied). Its quarter is a quarter of that field which has an evenly circular circumference (a genitive *tatpuruṣa*)[250].

When this is the interpretation (*vyākhyāna*) ⟨of the verse⟩ one ⟨wrongly⟩ arrives at an understanding of the area. This very analysis ⟨of the compound⟩ has been indicated by master Prabhākara. Because he is a *guru*, we are not blaming him.

Moreover: It is proper to say that a unit arc (*kāṣṭha*) can be equal (*tulya*) to its chord (*jyā*), even someone ignorant of treatises knows this; that a unit arc can be equal to its chord has been criticized by precisely this ⟨master⟩[251].

But we say: An arc equal to a chord exists. If an arc could not be equal to a chord then there would never be steadiness at all for an iron ball on level ground. Therefore, we infer that there is some spot by means of which that iron ball rests on level ground. And that spot is the ninety-sixth part of the circumference. That an arc can be equal to its chord, has been asserted even by other masters:

> **Because of its sphericity (*golakaśarīra*) ⟨a sphere⟩ touches the earth**
> **with the hundredth part of its circumference|**[252].

Samavṛttaparidhi is a circumference which is evenly circular (a *karmadhāraya*. *Samavṛttaparidhipada* is a **quarter of the circumference of an even circle** (a genitive *tatpuruṣa*). **One should divide** that quarter of the circumference of an even circle[253].

---

[249]This translation, as we have discussed in the supplement for this commentary, is in the line of Bhāskara's understanding of the verse. An explanation of Bhāskara's interpretation and the procedure he prescribes is given in this supplement.

[250]Bhāskara presents here Prabhākara's (probably an older commentator of Āryabhaṭa) interpretation of the compound. According to this understanding a quarter of a *disk* would be considered. Bhāskara, as we will see below, disagrees.

[251]This statement and the following discussion has been commented upon in the last section of the supplement for this commentary of verse.

[252]This whole paragraph is translated in the introduction of [Shukla 1976; 7.43.3, p. lxiv.]. And also in [Hayashi 1997; p.212-213].

[253]Contrasting with Prabhākara's interpretation, Bhāskara understands that this compound concerns a quarter of the circumference.

15   "By a partition (*vibhāga*) with chords" is the remaining part of the sentence. When the circumference of an even circle is being cut (*khaṇḍyamāna*) by a partition with chords, "**the half-chords of an even ⟨number of⟩ unit arcs**" are produced "**from a trilateral and a quadrilateral**", and not half-chords of an uneven ⟨number of⟩ unit arcs.

Just those particular ones, the numbers of which are doubled successively, are understood: two, four, eight, sixteen, thirty two, etc... And, due to the word "*tu*" (and), two, four, six, eight, ten, twelve, fourteen, etc...

"**On the semi-diameter**" a trilateral field is produced. The half-chords are produced from trilateral or quadrilateral fields[254].

p.78,
line 1   ⟨Question⟩

But how is produced on the semi-diameter a trilateral or a quadrilateral field?

It is replied: That ⟨field⟩ whose base (*bhujā*) or ear (*karṇa*) is the semi-diameter (*vyāsārdha*) is what is produced on the semi-diameter.

Or else, on that very semi-diameter (*viṣkambhārdha*[255]) the half-chords are produced. The meaning is: because they are a part of the semi-diameter, they do not occur beyond the semi-diameter.

Or else, when the semi-diameter exists, the half-chords are produced. Because when the semi-diameter is known one can determine the chords, and otherwise
5   not.

Why?

Because it has been stated: "The chord of one sixth of the circumference is equal to the semi-diameter." [Ab.2.9.cd].

Yatheṣṭāni is as many as one desires (*yathepsitāni*), half-chords of an even ⟨number of⟩ unit arcs.

In this verse (*kārikā*), only the essence of the production of chords (*jyotpatti*) has been put forth by the master but [the procedure] has not been given, because the procedure (*karaṇa*) is established in another place.

10   Or else, the procedure ⟨used⟩ in the production of chords relates to diagrams (*chedyaka*), and a diagram is intelligible (*gamya*) ⟨only⟩ with an explanation (*vyakhyāna*). Therefore it has [not] been put forth ⟨by Āryabhaṭa in the *Āryabhaṭīya*⟩.

⟨Question⟩

Now why is it only a quarter of the circumference of an even-circle that is divided by a partition with chords, and ⟨why⟩ is it not the ⟨entire⟩ circumference of the even-circle that is divided?

---

[254]Reading "*kṣetrāj jyārdhāni*" instead of "*kṣetrajyārdhāni*".

[255]From now on, unless otherwise stated, the compound translated by "semi-diameter" is *viṣkambhārdha*.

There is no mistake. The size of the quarter of the circumference of an even-circle is three *rāśis*. This is so in the four quarters. Because the sizes of the quarters of a circumference are equal ⟨to each other⟩, the half-chords of all quarters of the circumference are equal ⟨to each other⟩; therefore, only the half-chords of the quarter of a circumference [256] have been put forth in order to establish the practical computation (*vyavahāra*).

An example:

How much are the sizes of the half-chords on a semi-diameter measuring eight (*vasu*)- three (*dahana*)-four (*kṛta*)-three (*hutāśana*)?|

Unit arcs (*kāṣṭha*) which are half *rāśis* are produced. The semi-diameter is 3438.

Procedure: Having drawn a circle (*maṇḍala*) with a pair of compasses (*kakarṭa*) whose ⟨opening⟩ is equal to the semi-diameter determined by a size as large as ⟨desired⟩, one should divide that ⟨circle⟩ into twelve ⟨equal parts⟩. And these twelfth parts should be regarded as "*rā'sis*". Now, in the circle which is divided into twelve ⟨equal parts⟩, in the east one should make a line which has the form of a chord, and which penetrates (*avagāhinī*) ⟨the circle at⟩ the tips of two *rāśis* from south to north. Likewise in the western part also. In the same way, again, in the southern and northern parts also, one should make chords extending from east to west. And furthermore in the eastern, western, southern and northern directions, in exactly the same way, one should make lines which penetrate ⟨the circle at⟩ the tips of four *rāśis*. Then they should be made into trilaterals[257] ⟨by drawing the diagonals of the rectangles obtained⟩.

And thus a field produced by a circumference is drawn with a pair of compasses with a stick (*vartikā*) fastened to the opening (*mukha*)[258]. In the field drawn in this way all is to be shown.

In this drawing (*ālekhya* i.e., when the unit arc is half a *rāśi*) the [whole] chord of four unit arcs (*kāṣṭha*) is equal to the semi-diameter. Half of that is the ⟨half-⟩chord (*jyā*) of two unit arcs. And that is 1719.

This is the base (*bhujā*), the semi-diameter is the hypotenuse (*karṇa*), therefore the perpendicular (*avalambaka*) is the root of the difference of the squares of the base and the hypotenuse. That exactly is the ⟨half-⟩chord of four unit arcs. And that is 2978.[259] When one has subtracted this from the semi-diameter, the remainder is the arrow of ⟨the half-chord of⟩ two unit arcs. The hypotenuse is the root of the sum of the squares of the arrow and the ⟨half-⟩chord of two unit arcs. And that

---

[256] Reading *paridhipādajyārdhāny eva* rather than *paridhipādajyārdha ity eva*.

[257] Reading as in all manuscripts, *tryaśrīkartavyāni* rather than *tryaśrī[ṇi]* as suggested by the editor.

[258] We do not know why, having prescribed above a specific field, Bhāskara then indicates that another circle should be drawn.

[259] This result is an approximation. For more precision, please see the Annex to this commentary.

Figure 29:

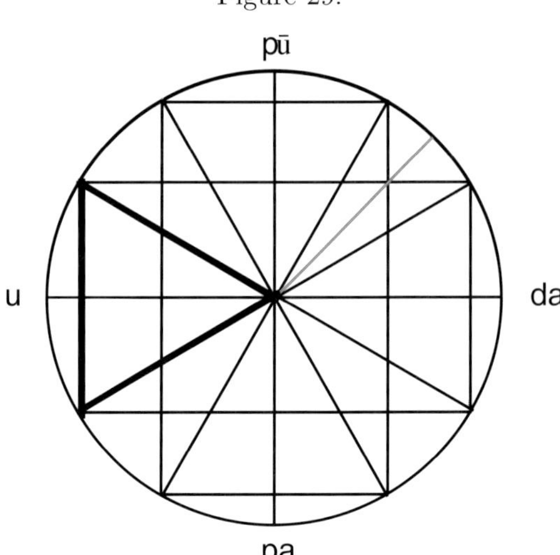

precisely is the [whole] chord of two unit arcs, which is 1780. Half of that is the ⟨half-⟩chord of one unit arc, 890.

This is the base, the semi-diameter is the hypotenuse. The perpendicular is the root of the difference of the squares of the base and the hypotenuse. And that is the ⟨half-⟩chord of five unit arcs. And that is 3321. Because ⟨the number, 5, of unit arcs⟩ is uneven, no ⟨half-⟩chords are produced from this. In this way, from a trilateral, five half-chords have been explained[260].

15   In the interior equi-quadrilateral field the sides are equal to the semi-diameter. Its diagonal (*karṇa*) is the root of the sum of the squares of two semi-diameters. And that is 4862. Its half is the ⟨half-⟩chord of three unit arcs. And that is 2431. In this way one ⟨half-⟩chord is produced from a quadrilateral; because ⟨the number of arcs it is related to⟩ is uneven, there is no production ⟨of further half-chords⟩.

p.80,
line 1   On the semi-diameter six half-chords whose unit arcs are half *rāśis* have been explained. On that very semi-diameter, we will explain the ⟨half-⟩chords whose unit arc is the fourth part of a *rāśi*. It is as follows:

In a field drawn as before the [whole] chord of eight unit arcs is exactly the semi-diameter. Half of that is the ⟨half-⟩chord of four unit arcs. And that is 1719.

This is the base, the semi-diameter is the hypotenuse; the upright side (*koṭi*) is the root of the difference of the squares of the base and the hypotenuse. That is

---

[260] Four half-chords have been computed, but the semi-diameter can be seen as the half-chord of twelve unit arcs, and may be assumed here. In the following, systematically, there is always one half-chord "missing" from Bhāskara's counting, and we may suppose that this is the one omitted.

the ⟨half-⟩chord of eight unit arcs, and that is 2978.

When one has subtracted this from the semi-diameter, the remainder is the arrow for the ⟨half-⟩chord of four unit arcs. The hypotenuse is the root of the sum of the squares of the arrow and the ⟨half-⟩chord of four unit arcs. That is the [whole] chord of four unit arcs, and that is 1780. Half of that is the ⟨half-⟩chord of two unit arcs, and that is 890.

This is the base, the semi-diameter is the hypotenuse, the upright side is the root of the difference of the squares of the base and the hypotenuse. That is the ⟨half-⟩chord of ten unit arcs, and that is 3321.

When one has subtracted this from the semi-diameter, the remainder is the arrow ⟨for the half-chord⟩ of two unit arcs. The hypotenuse is the root of the sum of the squares of the arrow and the ⟨half-⟩chord of two unit arcs. That precisely is the [whole] chord of two unit arcs, and that is 898. Its half is the ⟨half-⟩chord of one unit arc, and that is 449.

This is the base, the semi-diameter is the hypotenuse, the upright side is the root of the difference of the squares of the base and the hypotenuse. That is the ⟨half-⟩ chord of eleven unit arcs, and that is 3409. Because ⟨the number, 11, of unit arcs⟩ is uneven, no ⟨half-⟩chord is produced from this.

Now, when one has subtracted the ⟨half-⟩chord of two unit arcs from the semi-diameter, the remainder is the arrow ⟨for the half-chord⟩ of ten unit arcs. The hypotenuse is the root of the sum of the squares of the arrow and the ⟨half-⟩ chord of ten unit arcs. That [precisely] is the [whole] chord of ten unit arcs, and that is 4186. Its half is the ⟨half-⟩chord of five unit arcs, and that is 2093.

This is the base, the semi-diameter is the hypotenuse, the upright side is the root of the difference of the squares of the base and the hypotenuse. That is the ⟨half-⟩ chord of seven unit arcs, and that is 2728. Because ⟨the number,7, of unit arcs⟩ is uneven, no ⟨half-⟩chord is produced from this. In this way from a trilateral nine half-chords ⟨has been explained⟩.

As before, the diagonal of the mentioned equi-quadrilateral whose sides are semi-diameters, is the root of the sum of the squares of two semi-diameters. That is the [whole] chord of twelve unit arcs, and that is 4862. Its half is the ⟨half-⟩chord of six unit arcs, and that is 2431.

When one has subtracted this from the semi-diameter, the remainder is the arrow ⟨for the half-chord⟩ of six unit arcs, the hypotenuse is the root of the sum of the squares of the arrow and the ⟨half-⟩chord of six unit arcs. That precisely is the [whole] chord of six unit arcs, and that is 2630. Its half is the ⟨half-⟩chord of three unit arcs, and that is 1315.

That is the base, the semi-diameter is the hypotenuse, the upright side is the root of the difference of the squares of the base and the hypotenuse. That is the ⟨half-⟩chord of nine unit arcs, and that is 3177. Because ⟨the number, 9, of unit arcs⟩ is uneven, no ⟨half-⟩chord is produced from this. In this way three ⟨half-⟩

chords ⟨have been produced⟩ from a quadrilateral, ⟨and, in total⟩ twelve on the half diameter[261].

Twelve half-chords whose unit arcs are the fourth part of a *rāśi* have been explained. We will tell the ⟨half-⟩chord of the eighth part of a *rāśi* on that very semi-diameter. It is as follows:

In a field drawn as before the [whole] chord of sixteen unit arcs is precisely the semi-diameter. Half of that is the ⟨half-⟩chord of eight unit arcs, and that is 1719.

This is the base, the semi-diameter is the hypotenuse, the upright side is the root of the difference of the squares of the base and the hypotenuse. That is the ⟨half-⟩chord of sixteen unit arcs, and that is 2978.

One should subtract this from the semi-diameter. The remainder is the arrow ⟨for the half-chord⟩ of eight unit arcs. The hypotenuse is the root of the sum of the squares of the ⟨half-⟩chord of eight unit arcs and the arrow. That precisely is the [whole] chord of eight unit arcs, and that is 1780. Its half is the ⟨half-⟩chord of four unit arcs, and that is 890.

This is the base, the semi-diameter is the hypotenuse, the upright side is the root of the difference of the squares of the base and the hypotenuse. That precisely is the ⟨half-⟩chord of twenty unit arcs, and that is 3321.

When one has subtracted this from the semi-diameter, the remainder is the arrow ⟨for the half-chord⟩ of four unit arcs. The hypotenuse is the root of the sum of the ⟨half-⟩chord of four unit arcs and the arrow. Just that is the [whole] chord of four unit arcs, and that is 898. Its half is the ⟨half-⟩ chord of two unit arcs, and that is 449.

This is the base, the semi-diameter is the hypotenuse, the upright side is the root of the difference of the squares of the base and the hypotenuse. Just that is the ⟨half-⟩chord of twenty-two unit arcs, and that is 3409.

One should subtract this from the semi-diameter. The remainder is the arrow ⟨for the half-chord⟩ of two unit arcs. The hypotenuse is the root of the sum of the half-chord of two units arcs and the arrow. That precisely is the [whole] chord of two unit arcs, and that is 450. Its half is the ⟨half-⟩chord of one unit arc, and that is 225.

This is the base, the semi-diameter is the hypotenuse, the upright side is the root of the difference of the squares of the base and the hypotenuse. Just that is the ⟨half-⟩chord of twenty-three unit arcs, and that is 3431. Because ⟨the number, 23, of unit arcs⟩ is uneven, no ⟨half-⟩chord is produced from this.

Now, one should subtract the ⟨half-⟩chord of four unit arcs from the semi-diameter. The remainder is the arrow ⟨for the half-chord⟩ of twenty unit arcs. The hypotenuse is the root of the sum of the squares of the arrow and the ⟨half-⟩chord of twenty

---

[261]Once again, only 11 have been explicitly given.

unit arcs. That is the [whole] chord of twenty unit arcs, and that is 4186. Its half is the ⟨half-⟩chord of ten unit arcs, and that is 2093.

This is the base, the semi-diameter is the hypotenuse, the upright side is the root of the difference of the squares of the base and the hypotenuse. Just that is the ⟨half-⟩chord of fourteen unit arcs, and that is 2728 .

One should subtract this from the semi-diameter. The remainder is the arrow ⟨for the half-chord⟩ of ten unit arcs. The hypotenuse is the root of the sum of the squares of the ⟨half-⟩chord of ten unit arcs and its arrow. Just that is the [whole] chord of ten unit arcs, and that is 2210. Its half is the ⟨half-⟩chord of five unit arcs, and that is 1105.

This is the base, the semi-diameter is the hypotenuse, the upright side is the root of the difference of the squares of the base and the hypotenuse. That precisely is the ⟨half-⟩chord of nineteen (*ekonaviṃśati*) unit arcs, and that is 3256. Because ⟨the number, 19, of unit arcs⟩ has the state of being uneven, no ⟨half-⟩chord is produced from this.

Now one should subtract the ⟨half-⟩chord of two unit arcs from the semi-diameter. The remainder is the arrow ⟨for the half chord⟩ of twenty-two unit arcs. The hypotenuse is the root of the sum of the squares of the arrow and the ⟨half-⟩chord of twenty-two unit arcs. That precisely is the [whole] chord of the twenty-two unit arcs, and that is 4534. Its half is the ⟨half-⟩chord of eleven unit arcs, and that is 2267.

This is the base, the semi-diameter is the hypotenuse, the upright side is the root of the difference of the squares of the base and the hypotenuse. Just that is the ⟨half-⟩chord of thirteen unit arcs, and that is 2585. Because ⟨the number, 13, of unit arcs⟩ is uneven, no ⟨half-⟩chord is produced from this.

Now, one should subtract the ⟨half-⟩chord of ten unit arcs from the semi-diameter. The remainder is the arrow ⟨for the half-chord⟩ of fourteen unit arcs. The hypotenuse is the root of the sum of the squares of the ⟨half-⟩chord of fourteen unit arcs and the arrow. That, precisely, is the [whole] chord of fourteen unit arcs, and that is 3040. Its half is the ⟨half-⟩chord of seven unit arcs, and that is 1520.

This is the base, the semi-diameter is the hypotenuse, the upright side is the root of the difference of the squares of the base and the hypotenuse. That precisely is the ⟨half-⟩chord of seventeen unit arcs, and that is 3084. Because ⟨the number, 17, of unit arcs⟩ is uneven, no ⟨half-⟩chord is produced from this.

In this way, the ⟨half-⟩chords ⟨for the⟩ unit arcs which are the eighth part of a *rāśi* have been explained from trilaterals. Now we will explain ⟨the production of a half-chord⟩ from a quadrilateral. The sides of the interior equi-quadrilateral field are equal to the semi-diameter. The diagonal is the root of the sum of the squares of those two ⟨sides⟩. That, precisely, is the [whole] chord of twenty-four unit arcs, and that is 4862. Its half is the ⟨half-⟩chord of twelve unit arcs, and that is 2431.

One should subtract this from the semi-diameter. The remainder is the arrow ⟨for the half-chord⟩ of twelve unit arcs . The hypotenuse is the root of the sum of the squares of the arrow and the ⟨half-⟩chord of twelve unit arcs. That, precisely, is the [whole] chord of twelve unit arcs, and that is 2630. Its half is the ⟨half-⟩chord of six unit arcs, and that is 1315.

This is the base, the hypotenuse is the semi-diameter, the upright side is the root of the difference of the squares of the base and the hypotenuse. That is the ⟨half-⟩chord of eighteen unit arcs, and that is 3177.

One should subtract this from the semi-diameter. The remainder is the arrow ⟨for the half-chord⟩ of six unit arcs. The hypotenuse is the root of the sum of the squares of the ⟨half-⟩chord of six unit arcs and the arrow. That precisely is the [whole] chord of six unit arcs, and that is 1342.   Its half is the ⟨half-⟩chord of three unit arcs and that is 671.

This is the base, the semi-diameter is the hypotenuse, the upright side is the root of the difference of the squares of the base and the hypotenuse. Just that is the ⟨half-⟩chord of twenty-one unit arcs , and that is 3372. Because ⟨the number, 21, of unit arcs⟩ has the state of being uneven, no ⟨half-⟩chord is produced from this.

Now one should subtract the ⟨half-⟩chord of six unit arcs from the semi-diameter. The remainder is the arrow ⟨for the half- chord⟩ of eighteen unit arcs. The hypotenuse is the root of the sum of the squares of the arrow and the ⟨half-⟩ chord of eighteen unit arcs. Just that is the [whole] chord of eighteen unit arcs, and that is 3820.   Its half is the ⟨half-⟩chord of nine unit arcs, and that is 1910.

This is the base, the semi-diameter is the hypotenuse, the upright side is the root of the difference of the squares of the base and the hypotenuse. That precisely is the ⟨half-⟩chord of fifteen unit arcs, and that is 2859. Because ⟨the number, 15, of unit arcs⟩ is uneven, no ⟨half-⟩chord is produced from this.

In this way twenty-four ⟨half-⟩chords ⟨subtended by⟩ unit arcs which are the eighth part of a *rāśi* ⟨have been explained⟩. With this very method (*vidhāna*) on a semi-diameter as many half ⟨half-⟩chords as one desires are to be produced.

[Segmented ⟨half-⟩chords with a different method]

Now, in order to teach a partition of ⟨half-⟩chords (*jyāvibhāga*), he states:

12.[262] The segmented second half-⟨chord⟩ is smaller than the first half-chord of a ⟨unit⟩ arc by certain ⟨amounts⟩ |

---

[262]This is a translation of the verse according to Bhāskara's understanding of it. Historians of mathematics consider it a misinterpretation of Āryabhaṭa's verse. See [Hayashi 1997] and the Annex to this verse.

> The remaining ⟨segmented half-chords⟩ are smaller ⟨than the first half-chord, successively⟩ by those ⟨amounts⟩ and by fractions of the first half-chord accumulated.‖

prathamāc cāpajyārdhāt yair ūnaṃ khaṇḍitaṃ dvitīyārdham|
tatprathamajyārdhāṃśais tais tair ūnāni śeṣāni‖

Prathamāt is **than the first** (*ādyāt*) "**half-chord of a ⟨unit⟩ arc**".

⟨As for⟩ "**is smaller by certain ⟨amounts⟩**", ⟨it⟩ is smaller, that is shorter, by certain parts (*aṃśa*)[263].

What is that ⟨which is smaller than the first half-chord⟩?

"**The segmented second half ⟨chord⟩**" , ⟨it⟩ is segmented, ⟨in other words⟩ the second half-chord of ⟨unit⟩ arc is cut (*chinna*) with the diagrammatical rule (*chedyakavidhi*) which has been mentioned in the previous *āryā* ⟨verse⟩.

"**By those ⟨amounts⟩ and by fractions of the first half-chord**"; with ⟨the word⟩ "those", they[264] understand so many ⟨amounts⟩, by means of which the second half-chord of ⟨unit⟩ arcs is less than the first half-chord of a ⟨unit⟩ arc.

Jyārdha is half of the chord (a genetive *tatpuruṣa*), and *prathamajyārdha* is that which is first and half of the chord (a *karmadhāraya*). Alternatively, *prathamajyā* is that which is first and which is a chord (a *karmadhāraya*), *prathamajyārdha* is that which is the first chord and which is a half (a *karmadhāraya*)[265].

Prathamajyārdhāṃ'sa is a fraction of the first half-chord (a genitive *tatpuruṣa*).

And a fraction of the first half-chord is what has been obtained when one has divided (*bhāgaṃ hṛtvā*) by the first half-chord, just like "a fraction of five" (one fifth) and "a fraction of six" (one sixth)[266]. *Tatprathamajyārdhāṃśa* is those ⟨amounts⟩ and the fractions of the first half-chord (a *dvandva*). By those ⟨amounts⟩ and the fractions of the first half-chord. The expression *tais tair aṃśair* stated with a repetition has the meaning of "*ca*" (accumulation).

⟨As for⟩ "**the remaining ⟨half-chords⟩ are smaller**". "Smaller", in other words "deprived of" (*rahita*), "the remaining" that is, the third and following ⟨partial⟩ half-chords.

It is as follows: This first half-chord of a ⟨unit⟩ arc produced with a diagram is 225. The second partial (*cheda*) half-chord of ⟨unit⟩ arcs is 224. This is smaller than the first half-chord of ⟨unit⟩ arc by one. The sum (*ekatra*) of the second partial (*aṃśa*[267]) half-chord of ⟨unit⟩ arc and of the first half-chord is 449. When

---

[263]For an explanation of the mathematical content referred to here and below please see the supplement for this verse

[264]"They" may be a reference to the followers of Prabhākara, according to the statement below, [Shukla 1976; p. 84 line 13].

[265]The compound has been translated as: "the first half-chord".

[266]The compound *prathamajyārdhāṃ'sa* ends with the word *aṃśa*, translated by "fraction", this word is also used as a fraction marker: one fifth is expressed by the compound *pañcāṃ'sa*.

[267]From here on the word *aṃśa*, understood as a synonym of *khaṇḍita* will be translated by "partial".

the division of this with the first half-chord ⟨is made⟩, the quotient, because it is greater than one half, is two unities. [The third ⟨partial⟩ half-chord of unit arc is smaller] than the first half-chord of ⟨unit⟩ arc [by one] and by these previous ⟨two⟩. And that is 222. The sum (*saṃyoga*) of the three ⟨partial half-chords⟩ is 671. The division of that with the first half-chord of a ⟨unit⟩ arc ⟨is made⟩, the quotient, because it is greater than one half, is three unities. The fourth ⟨partial⟩ half-chord is smaller than the first half-chord of ⟨unit⟩ arc, by these ⟨three⟩ and by the previously obtained fractions, and that is 219. The sum of the four ⟨partial⟩ half-chords is 890, the division of that with the first half-chord ⟨is made⟩, the quotient, because it is greater than one half, is 4 unities. The fifth half-chord of a ⟨unit⟩ arc is smaller than the first half-chord of ⟨unit⟩ arc by six and by those previous fractions (*aṃ'sa*). And that is 215. With these ⟨words⟩ the remaining ⟨partial half-chords⟩ have ⟨also⟩ been explained.

And this interpretation has been given by master Prabhākara. It is improper to give an explaination without having rejected what is useless.

⟨Question⟩

Why is it useless?

In this treatise on mathematics a different rule (*sūtrāntara*) is put forth in order to teach a different method (*upāyāntara*) or ⟨otherwise⟩ in order to teach an easier method (*laghūpāya*). In this case there is no scent (*gandha*, i.e. hint) of either one ⟨of these reasons⟩.

⟨Question⟩

Why?

This computation (*karman*) is made with the first and second ⟨partial?⟩half-chords of ⟨unit⟩ arc which are definitely known by means of the diagrammatic method stated in the previous *āryā* ⟨verse⟩. In this case the method is not easy because it depends on two rules. And ⟨it is not⟩ a different method because it depends on the previous rule. Therefore, no ⟨new⟩ purpose ⟨is obtained⟩ with this rule.

⟨Question⟩

But how are these respective ⟨partial half-⟩chords known?

These are childish words! It is ⟨known⟩ from the production of ⟨half-⟩chords (*jyot-patti*). The half-chords of one ⟨unit⟩ arc (*kāṣṭha*), two ⟨unit⟩ arcs, three ⟨unit⟩ arcs etc., are explained ⟨under the previous verse⟩. Even someone ignorant of mathematics knows that ⟨partial half-⟩chords are ⟨obtained⟩ one by one by ⟨taking⟩ their mutual difference. Much more a calendar maker (*sāṃvatsaraḥ*)! And thus it is assembled in order to bestow intelligence to the dull-minded one. As follows:

225, 449, 671, 890, 1105, 1315, 1520, 1719, 1910, 2093, 2267, 2431, 2585, 2728, 2859, 2978, 3084, 3177, 3256, 3321, 3372, 3409, 3431, 3437.[268]

---

[268] We can recognize here an enumeration of the Rsinuses given by Bhāskara in his commentary to verse 11 of the chapter on mathematics, in his last example of use of the procedure.

Decreased by the immediate ⟨predecessors⟩ one by one, ⟨they are⟩ in order the ⟨partial half-⟩ chords:

225, 224, 222, 219, 215, 210, 205, 199, 191 , 183, 174, 164, 154, 143, 131, 119, 106,   10
13, 79, 65, 51, 37, 22, 7.[269]

These ⟨partial half-chords, added⟩ in the reverse order beginning from the last, are the *utkramajyā* (Rversed sine).

[Correct production of a circle and so on]

Now, in order to indicate the mere direction (*diṅmātra*) of what has not been taught ⟨previously⟩[270], he states:

> **Ab.2.13. A circle should be brought about with a pair of compasses, and a trilateral and a quadrilateral each ⟨are brought about⟩ with two diagonals|**
> **Flat ground should be brought about with water, vertically (literally: top and bottom) with just a plumb-line‖**
>
> **vṛttaṃ bhrameṇa sādhyaṃ tribhujaṃ ca caturbhujaṃ ca karṇābhyām|**
> **sādhyā jalena samabhūr adho ūrdhvaṃ lambakenaiva‖**

A circular field is brought about with a *bhrama*. With the word *bhrama* a pair of compasses (*karkaṭa*) is understood. With that pair of compasses an evenly circular field is delimited by the size of the out-line (*parilekhā*).

⟨As for:⟩ "**Both a trilateral and a quadrilateral with diagonals**". A trilateral field and a quadrilateral field should each be brought about with two diagonals. First a trilateral:

Having stretched a string (*sutra*) on level ground one should make a line (*rekha*). And that is: 20

Figure 30:

____

Here, with a pair of compasses (*karkaṭaka*) which is placed on both tips ⟨of the line⟩, a fish should be produced.

---

[269]This is the enumeration of sine differences given in Ab.1.12, see [Sharma-Shukla 1976; p.29].
[270]All manuscripts read, according to note 4 p. 84: *anādiṣṭa*, i.e. "what has not been taught". The main text of the edition reads: *vṛttādisiddhiṃ*, i.e. "the correct production of a circle, etc."

p.85, line 13
p.86, line 1

A perpendicular is a second string which goes from the mouth to the tail of this ⟨fish⟩:

Figure 31:

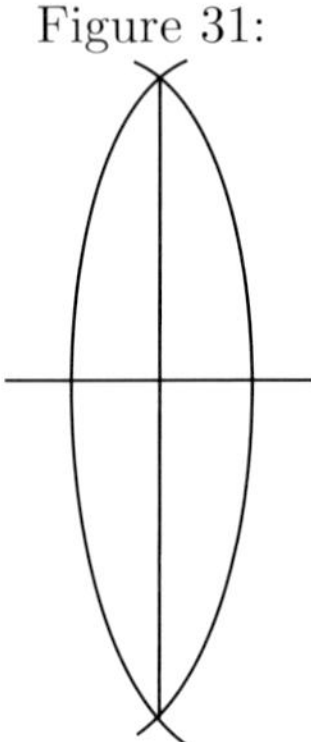

5

Having appointed one tip of a string on the extremity ⟨of the fish⟩, having appointed the second tip ⟨of the string⟩ firmly on the tip of the base, one should make a line. On the second tip ⟨of the base⟩, too, it is just in that way. In this way, there are two diagonal strings. With those two diagonal strings a trilateral is brought about:

Figure 32:

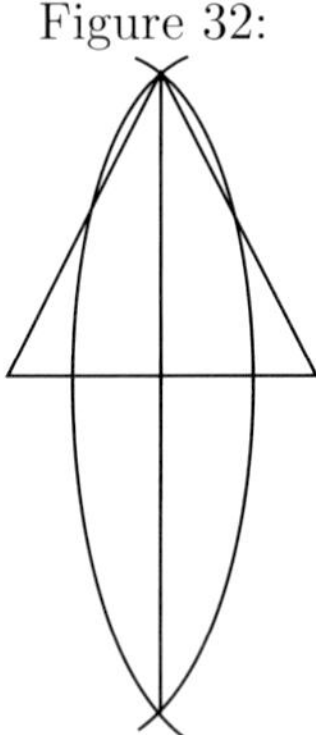

In ⟨the case of⟩ a quadrilateral, one should stretch obliquely a string which is equal
to [the diagonal of] the desired quadrilateral. And that string is:

Figure 33:

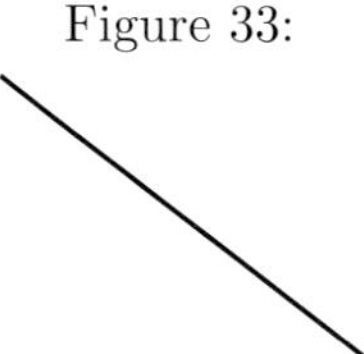

One should stretch obliquely the second ⟨string⟩ too, a cross (*svastika*) is produced
from the middle of that ⟨first string⟩. And therefore there are two diagonal strings:

Figure 34:

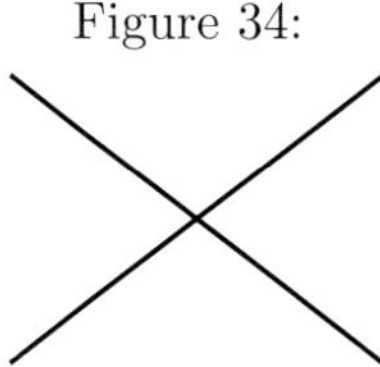

The sides (*pārśva*) of these two ⟨strings⟩ are filled in, ⟨and⟩ a quadrilateral field is
produced:

Figure 35:

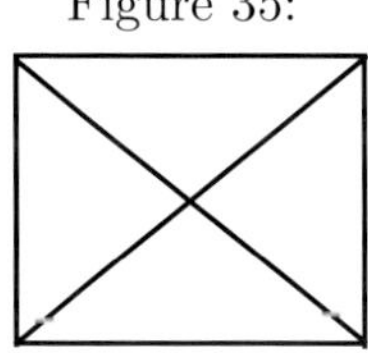

⟨As for:⟩ **"Flat ground should be brought about with water"**. Flat ground is p.87,
line 1
brought about (*sādhyate*) with water. It is as follows:

Having set down a jar of water, when there is no wind, above a triple timber, on
a ground ⟨provisionally⟩ levelled with the eye[271], one should make a hole at the
bottom ⟨of the jar⟩, in such a way that the water flows in the form of a uniform
stream. Where the water flowing from it spreads in a circle on every side, the
ground (*bhū*) is flat; where that water having broken the circle, stands still, ⟨the
ground⟩ is depressed; where it cannot enter,⟨the ground⟩ is elevated.

---

[271]Lit. "with eye-rays" or "eye-strings" (*cakṣuḥsūtra*).

⟨As for:⟩ **"verticality with just a plumb-line"**. The point above a ⟨point⟩ marked below is brought about with a plumb-line (*avalambaka*). Or, the point below a point above, also, is ⟨brought about⟩ with a plumb-line. And a plumb-line is a thread with a heavy object at one tip.

[The semi-diameter of one's own circle]

p.87,
line 9 In order to compute the semi-diameter (*viṣkambhārdha*) of one's own circle (*sva-vṛtta*), he states:

> **Ab.2.14. Having summed the square of the size of a gnomon and the**
> **square of the shadow |**
> **The square root of that ⟨sum⟩ is the semi-diameter of one's own circle‖**
>
> **śaṅkoḥ pramāṇavargaṃ chāyāvargeṇa saṃyutaṃ kṛtvā|**
> **yat tasya vargamūlaṃ viṣkambhārdhaṃ svavṛttasya‖**

[Discussing the different shapes of gnomons]

Here, calendar makers (*sāṃvatsara*) disagree on the size and shapes of the gnomon (*śaṅku*).

First, some say: "A gnomon of twelve *aṅgulas* has four edges (*caturaśra*) on ⟨its⟩ lower (*mūla*) third, has three edges (*tryaśri*) on ⟨its⟩ middle (*madhya*) third and has ⟨the form of⟩ a spear on its upper third. Because ⟨the top of the gnomon⟩ has a sharp (*sūkṣma*) shape and because it is easy to characterize (*lakṣaṇa*) a shadow by means of one sharp upright side (*koṭi*) and because it is difficult to acquire by all other ⟨means, this is a good gnomon⟩".

This is not so. Because it is impossible to arrange a plumb-line (*avalambaka*) whose tip ⟨is at the extremity of the part which has the form of⟩ a spear, verticality (*ṛjuta*) itself will be difficult to acquire. Because of that impossibility, all ⟨its⟩ qualities disappear. And, ⟨a plumb-line⟩, which has a circular belly, having the form of a cow's tail is rejected[272] precisely because ⟨it⟩ is a flawed plumb-line.

Others say: " ⟨It should⟩ have four edges because it is possible to bring about and secure with a plumb-line (*avalambaka*) four directions[273], and, the knowledge of the direction ⟨of the sun⟩ is established, in the direction of any desired upright side, from the knowledge of the shadow, with two ⟨opposite⟩ upright sides".

---

[272]Reading *pratyākhyataḥ* rather than *pratyākhyatā* as in the printed edition.

[273]Reading *caturdiśām* rather than the *caturdiśam* of the printed edition.

This also is proper; however, such ⟨an instrument⟩ is ⟨difficult to obtain⟩ at once because an artisan (*śilpin*) who produces a uniform four edged field (*samacaturaśrakṣetra*) is difficult to find. Even if someone with well exercised skill may sometimes be available, then also because ⟨it should⟩ be placed at every moment facing the sun, constantly the face (*mukha*) of the gnomon should be made to move. But since, then, for ⟨that gnomon⟩ which at first seems exceedingly precise (*atisūkṣma*), the desired shadow will be slightly exceeding, ⟨there is⟩ a draw-back (*doṣa*). Therefore, this gnomon also should be set aside. Gnomons are used everywhere with this very ⟨form⟩.

The followers of Āryabhaṭa, wishing to ground firmly their own thoughts, describe ⟨a gnomon⟩ as follows:

The best ⟨gnomon⟩ indeed is made of excellent wood, has no holes, is without streaks, knots or fractures; is produced (*siddha*) with a pair of compasses (*bhrama*), has the shape of a circle which is the same at the base, the middle, the top, and in the intermediate space; has a big diameter (*vyāsa*) and a big length (*ayāma*). Its vertical position (*ṛjusthiti*) is to be secured with three or four plumb lines.

Because the middle thread (*madhyasūtra*) of the gnomon has not been correctly established, its perpendicular position (*avalambakasthiti*) as well will be difficult to acquire. Therefore the securing of the middle thread of the gnomon is shown. It is as follows:

When one has placed the gnomon, firmly, on an elevated spot, having found the two middle points of the gnomon's base and top respectively, and having extended a thread fixed to its tip, one should make two lines on each side (*parśva*). These are the two middle lines (*madhyalekha*) on each pair of sides; then, once again, having produced, with a pair of iron compasses (*karkaṭa*), a fish from the two middle threads ⟨which went through⟩ the base and the top, one secures the remaining ⟨two⟩ middle lines.

⟨Objection⟩

But here also, there is decidedly a drawback (*doṣa*). In all directions, because the tip of the shadow of the top of that ⟨gnomon⟩, is a broad circle, the middle of the shadow is difficult to characterize; and in its absence there is no knowledge of the beginning ⟨of the shadow⟩.

This is no drawback. At the center (*kendra*) on top of the gnomon, another evenly circular stick, whose ⟨height⟩ surpasses the semi-diameter ⟨of the supporting cylinder⟩, made of metal or wood, as an ornament for the center, is made. Then the knowledge of the beginning and the finding of the middle ⟨line of the shadow⟩ which is the knowledge[274] of the beginning ⟨of the shadow⟩ will also be accomplished.

Alternatively, a wise man knows the middle ⟨line of the shadow⟩, using the previously fixed perpendicular string raised upwards a bit.

---

[274]Reading: *ādigrahaṇaṃ madhyaparijñāṃ ca'* instead of what is in the printed edition.

20 Now, because of the partition in *aṅgulas*, slightly, with a sharp knife, indents are made. Indeed, otherwise mentioning the size ⟨in the verse commented⟩ would be useless. Consequently, a gnomon which has the size one wishes when one has accepted a popular ⟨size⟩ is ⟨usually⟩ told to be "twelve *aṅgulas*". We will explain this in examples. The more that ⟨gnomon⟩ is wide and heavy, the less it will be moved by the wind; and the higher the parts in *aṅgulas* will be, the more precise (*sūkṣma*) ⟨compared to the length of the gnomon⟩ and well known ⟨it will be⟩. Therefore one should pay attention to ⟨gnomoms⟩ that are wide, heavy and tall. Thus the form of the ⟨desirable⟩ gnomon has been stated.

[Discussion on the size of gnomons]

Now we will explain the size (*pramāṇa*). Some have said: "⟨A gnomon⟩ is a half *hasta*[275] ⟨in length⟩ and has a body divided in twelve ⟨equal⟩ parts".

p.89, line 1 This is not a rule (*niyama*). However, the meaning is: the body divided (*pravibhakta*) with any desired number is divided with any desired number[276]. When the size is known, then again skill should be acquired, in order to have a partition of ⟨the gnomon⟩ in equal *aṅgulas* or when dividing into two at the center.

[Explanation of the verse (*śloka*)]

p.89; 5 ⟨As for⟩:"The square of the size of the gnomon"; this has been stated in order to explain the unsettled size of the stated "size of the gnomon", whose size has thus been fully explained ⟨before⟩. If the size of the gnomon had been precisely settled then, when stating this much, "the square of the gnomon", precisely that settled lsize is acknowledged also. *Pramāṇavarga* is "**the square of the size**" (a genetive *tatpuruṣa*).

⟨As for⟩ "**with the square of the shadow**". *Chāyāvarga* is the square of the shadow (genetive *tatpuruṣa*). "**Having summed**", the meaning is: having made into one (*ekīkṛtya*). "That which is the square root of that", that which is the square root of that summed quantity, that is the semi-diameter of one's own circle (the correlative of *yat* – the marker of a subordinate clause – is not expressed in the verse; *tasya* refers to the sum, and not to *yat*).

⟨Question⟩

What is that circle which has this semi-diameter?

It is replied:

It is the semi-diameter of that circle which is drawn with a pair of compasses (*karkaṭa*) ⟨whose opening is⟩ equal to the square root of that ⟨sum⟩.

---

[275]Half a *hasta* is twelve *aṅgulas*. For more details please refer to the section of the Glossary on Measuring Units.

[276]This seems to be a meaningless tautology, probably due to a corruption of manuscripts.

⟨Objection⟩

If so, every particular number indeed produces the semi-diameter of one's own circle.

There is no mistake. If every particular number becomes the semi-diameter of one's own circle, then what is disturbed by that ⟨question⟩[277]? Furthermore in this case just a particular semi-diameter of one's own ⟨circle⟩, which is the root of the sum of squares of the shadow and the size of the gnomon, is understood. Therefore, the understanding of another semi-diameter of one's own circle is never possible in this case. And when it is possible, a refutation (*parihāra*) of the mistake is prescribed.

And in this case, the stating of a semi-diameter of one's own circle is ⟨made⟩ in order to establish a Rule of Three: "If for the semi-diameter of one's own circle both the gnomon and the shadow ⟨have been obtained⟩, then for the semi-diameter of the ⟨celestial⟩ sphere, what are the two ⟨quantities obtained⟩?" In that way are obtained the Rsine of altitude (*śaṅku*) and the Rsine of the zenith distance (*chāyā*). Precisely, these two on an equinoctial day are told to be the Rsine of colatitude (*avalambaka*) and the Rsine of the latitude (*akṣajyā*).

An example:

1. The shadow of a gnomon divided into twelve is seen on level ground|
To be ⟨respectively⟩ five, nine and three and a half (lit. of which the
fourth is a half, *ardhacathurthā*) for the equinoctial mid-⟨day⟩ sun ‖

Setting down: the gnomon is 12, the shadow 5; the gnomon is 12, the shadow 9; the gnomon is 12, the shadow $\begin{vmatrix} 3 \\ 1 \\ 2 \end{vmatrix}$.

Procedure: The squares of respectively the gnomon and the shadow, 144, 25 are summed, 169. The root of this is the semi-diameter of one's own circle, and that is this: 13. The setting down for this field is:

The thread starting from the tip of the shadow and reaching the top of the gnomon is called "the semi-diameter of one's own circle". When one has set down the eye, along that thread, on the earth, one sees the sun adhering to the top of the gnomon.

When computing the Rsine of latitude (*akṣajyā*) a Rule of Three is set down: 13, 5, 3438. What is obtained is the Rsine of latitude, 1322[278]. That is the base (*bhujā*), the semi-diameter is the hypotenuse (*karṇa*); the root of the difference of the squares of the base and the hypotenuse is the Rsine of colatitude (*avalambaka*), 3174[279].

---

[277]Reading *tacchāyā* as in all manuscripts rather than *naśchāyā* of the printed edition.

[278]This is an approximate value. If we take, as we will systematically in the following footnotes, an approximation of a 10^{-2} range, then $1322 \simeq 1322,31$.

[279]This value is an approximation: $3174 \simeq 3173,67$.

Figure 36:

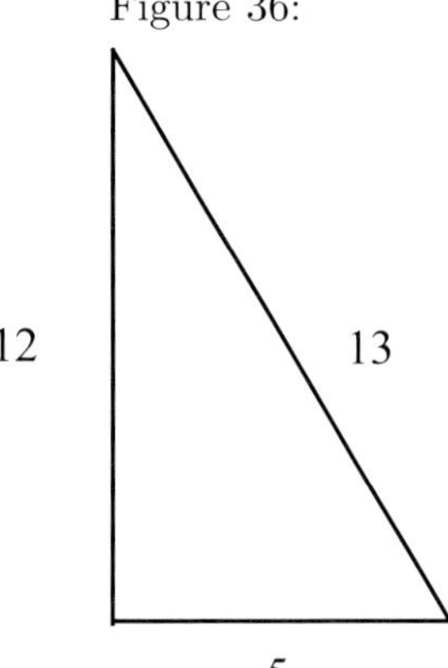

With a Rule of Three also 13, 12, 3438; what has been obtained is the Rsine of the colatitude, 3174[280]. There are other particular different fields, too, ⟨in this example⟩[281].

The verbal formulation (*vāco yukti*) of the Rule of Three is: "If for that semi-diameter of one's own circle, the base is equal to the shadow, the perpendicular is equal to the ⟨height of the⟩ gnomon, then what are both the base and the perpendicular for the semi-diameter (*vyāsa*) of the ⟨celestial⟩ sphere"?

When computing the ⟨time in⟩ *ghaṭikās* by means of the shadow and when computing the ⟨altitude of the⟩ sun by means of the midday shadow, just that method (*vidhi*) ⟨is used⟩ for the semi-diameter of one's own circle. However, ⟨this has been told⟩: when computing the ⟨time in⟩ *ghaṭikās* by means of the shadow, ⟨this⟩ should be performed with the Rsine of altitude (*śaṅku*); then just the gnomon (*śaṅku*) is computed. When computing the ⟨zenith distance of the⟩ sun with the shadow of ⟨the sun when it is on⟩ the prime vertical (*samamaṇḍala* i.e. at mid-day), ⟨it is⟩ just like that; when computing the sun with the midday shadow, the Rsine of the zenith distance (*natajyā*) is needed, in this way (*iti*) the shadow (*chāyā*) is computed.

For the two remaining ⟨fields⟩ also the semi-diameters of one's own circle are 15,
12
1 .
2

With just a Rule of Three the Rsine of latitude and the Rsine of colatitude are 2063, 2750; 963, 3300[282].

---

[280]This value is an approximation: $3174 \simeq 3173, 54$
[281]We do not know to which fields Bhāskara is referring here.
[282]The exact values found here are respectively: 2062,8 and 2750,4; 962,64 and 3300,48.

An example: 15

> **2. The shadow, for a gnomon of fifteen *aṅgulas*, is six *aṅgulas* and one**
>   **fourth |**
> **At midday on an equinox. Say in this case the Rsine of latitude and the**
>   **Rsine of colatitude. ‖**

Setting down: the gnomon is 15, the shadow $\begin{vmatrix} 6 \\ 1 \\ 4 \end{vmatrix}$. What is obtained is the semi-

diameter of one's own circle: $\begin{vmatrix} 16 \\ 1 \\ 4 \end{vmatrix}$. With this semi-diameter of one's own circle

are obtained the Rsine of latitude and the Rsine of colatitude, 1322, 3174[283].

An example: l. 20

> **3. When the shadow of a gnomon of thirty *aṅgulas* is seen as sixteen**
>   ***aṅgulas* |**
> **How much the sun, whose rays are spread out, has gone from the zenith**
>   **(*madhya*) should be then told. ‖**

Setting down: the gnomon is 30, the shadow is 16. What is obtained is the semi-
diameter of one's own circle: 34. What is obtained is the Rsine of the latitude of
that ⟨locality⟩ 1618[284].

[Computation of the shadow of a light]

                                                                              p.90; l.
                                                                              25
He states the computation (*karman*) of the shadow of a light (*pradīpa*):

> **Ab.2.15. The distance between the gnomon and the base, multiplied**
>   **by the ⟨height of⟩ the gnomon, is divided by the difference of the**
>   **⟨heights of the⟩ gnomon and the base.|**
> **What has been obtained should be known as that shadow of the gnomon**
>   **⟨measured⟩ indeed from its foot.‖**

śaṅkuguṇaṃ śaṅkubhujāvivaraṃ śaṅkubhujayor viśeṣahṛtam|
yal labdaṃ sā chāyā jñeyā śaṅkoḥ svamūlāt hi‖

---

[283]The values obtained here are approximations: $1322 \simeq 1322, 30$ and $3174 \simeq 3173, 54$.
[284]This value is an approximation: $1618 \simeq 1617, 88$.

Śaṅkuguṇa is that which has the ⟨height of⟩ the gnomon for multiplier (a *bahuvrīhi* compound translated as: "multiplied by ⟨height of⟩ the gnomon"). What is that ⟨which has the gnomon for multiplier⟩? he says: "**The distance between the gnomon and the base**". The word *bhujā* expresses the height (*ucchrāya*) of the light. The space (*antarāla*) between the height of the light and the gnomon is the distance between the gnomon and the base, that is multiplied by the ⟨height of⟩ the gnomon.

⟨As for⟩ "**is divided by the difference of the ⟨heights of⟩ the gnomon and the base**", *śaṅkubhujāvivara* is the difference of the ⟨heights of⟩ the gnomon and the base (a genitive *tatpuruṣa* with a sub-*dvandva*), it is *hṛta*, ⟨in other words⟩ divided (*bhakta*) by that.

⟨As for⟩ "**what has been obtained is that shadow of the gnomon**", ⟨measured⟩ precisely from that ⟨shadow's⟩ own foot, ⟨in other words:⟩ the shadow ⟨measured⟩ from the foot of just that gnomon has been obtained.

An example:

> **1. Say the shadows of the two gnomons which stand respectively at eighty from the foot of a light on a pole, whose height is seventy-two|**
> **And at twenty from ⟨a light whose height is⟩ thirty ‖**[285]

Setting down:

Figure 37:

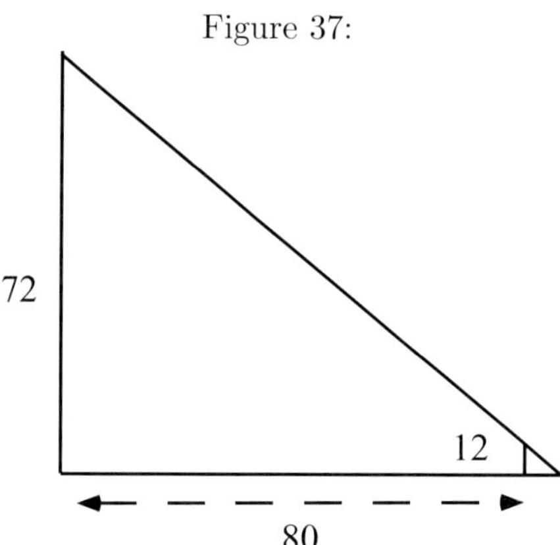

The distance between the gnomon and the base is 80; this, multiplied to the ⟨height of the⟩ gnomon is 960. The base is 72, the gnomon 12, their difference is 60; the distance between the gnomon and the base multiplied by the ⟨height of the⟩ gnomon is divided by this; the shadow is obtained: 16.

---

[285]It is understood, as it has been discussed in BAB.2.14, that the length of the gnomon is twelve *aṅgulas*.

Setting down the second example:

Figure 38:

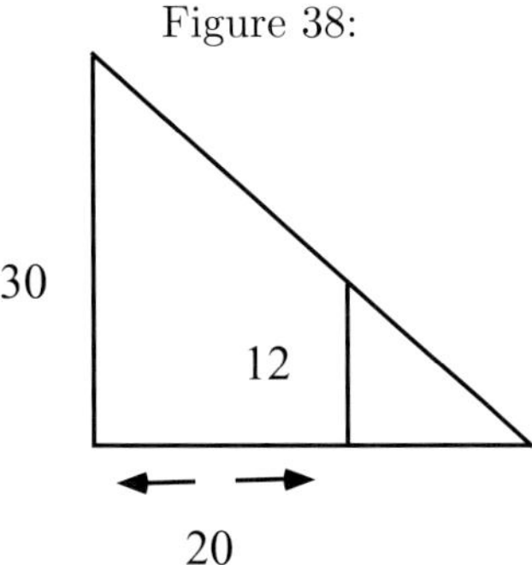

With the same procedure as before the shadow obtained is $13\frac{1}{3}$. This computa- p.92, line 1 tion (*karman*) is a Rule of Three. How? If from the top of the base which is greater than the gnomon, the size of the space between the gnomon and the base, which is a shadow, has been obtained, then, what is ⟨obtained⟩ with the gnomon? The shadow is obtained.

An example with a reversed operation (*viparikarman*):

p.92; l. 5

> **2. A shadow is seen as sixteen for a light whose height is seventy-two |**
> **How much is ⟨the distance⟩ to the foot of a gnomon whose ⟨height⟩ is**
> **twelve? ⟨This⟩ should be said by you. ‖**

Setting down:

Figure 39:

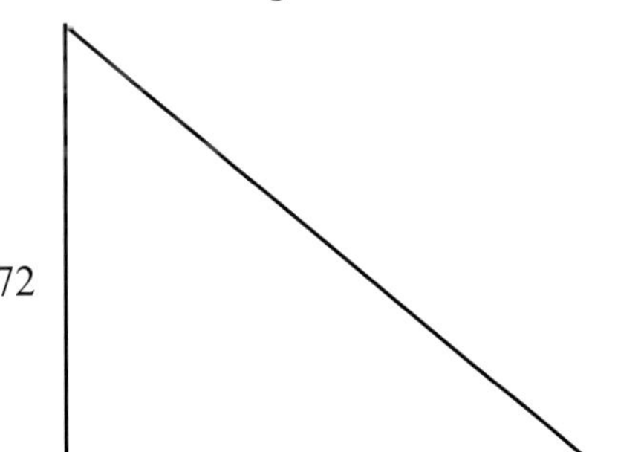

Procedure: With that difference between the ⟨the height of⟩ the gnomon and the base, 60, the shadow has been obtained. With this ⟨rule⟩, "the divisors become multipliers " ([Ab.2.28]), the shadow, 16, is multiplied by that, what results is 960. Just this is "the distance between the gnomon and the base having the ⟨height of the⟩ gnomon for multiplier". Here also the gnomon was a multiplier, ⟨therefore⟩ according to: "the multipliers become divisors" ([Ab.2.28]) , ⟨the previous quantity⟩ is divided by the ⟨height of the⟩ gnomon, 12, the distance between the gnomon and the base has been obtained, and that is 80.

An example:

> **3. A gnomon stands at a distance of fifty from the foot of a light on a
>   pole|**
> **Its shadow is ten (*pankti*). What should be stated is how much is ⟨the
>   height of⟩ the light in that ⟨case⟩ ‖**

Setting down :

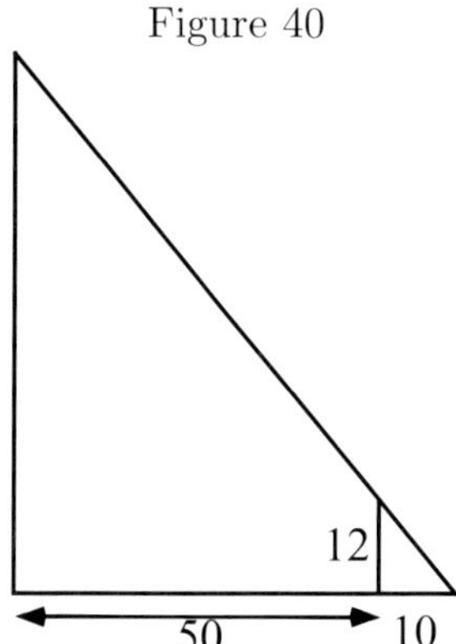

Figure 40

Procedure: With a procedure that will be said:

> "That upright side, having the gnomon for multiplier, divided by ⟨its⟩
> shadow, becomes the base ‖"[Ab.2.16.cd].

The shadow increased by the space between the base and the gnomon is the upright side. Therefore, the space between the gnomon and the base, 50, and the shadow, 10, are summed: 60. This multiplied by the ⟨height of the⟩ gnomon is 720. Divided by the shadow, it is the size of the base, 72.

[Knowing the distance and the height of a light with the shadows of two gnomons]

He states the computation of the unknown distance (*avasāna*) and height (*ucchrāya*) of a light with the shadows of two gnomons[286]:

> **Ab.2.16. The upright side is the distance between the tips of the ⟨two⟩ shadows multiplied by a shadow divided by the decrease.|**
> **That upright side multiplied by the gnomon, divided by ⟨its⟩ shadow, becomes the base ‖**
>
> **chāyāguṇitaṃ chāyāgravivaram ūnena bhājitaṃ koṭī|**
> **śaṅkuguṇā koṭī sā chāyābhaktā bhujā bhavati‖**

Chāyāguṇita is "**multiplied by the shadow**" (an instrumental *tatpuruṣa*). What is multiplied by the shadow? "**The distance between the tips of the shadows**". 10
Chāyāgravivara is the distance between the tips of the shadows (a dual genitive *tatpuruṣa*); the meaning is: the earth (*bhūmi*) in the space between the tips of the two shadows.

It is as follows: A gnomon has been fixed at some distance from a known light on a pillar of unknown height. Its shadow's ⟨length⟩ is known indeed. At a ⟨certain⟩ distance measured from the tip of its shadow there is a second gnomon. The distance (*vivara*) between the tips of the shadows is the distance (*antara*) defined as (*iti*) the tip of the shadow of the previous gnomon ⟨as measured⟩ from the tip of the shadow of that ⟨second gnomon⟩. It is multiplied, by the desired ⟨shadow⟩, the first shadow or the second shadow.

⟨As for⟩: "divided by the decrease" , the decrease (*ūna*), is the difference (*viśeṣa*) of 15
the shadows; it is divided by that decrease. The upright side is the earth (*bhūmi*) within the boundary (*avasāna*) ⟨delimited by the foot of the light and the tip of the shadow⟩. If that ⟨difference⟩ is multiplied by the first shadow, then ⟨the result of the computation⟩ becomes the space between ⟨the foot of⟩ the light on a pillar and the tip of the first gnomon⟨'s shadow⟩. If that ⟨difference⟩ is multiplied by the second shadow, ⟨then the result becomes⟩ the space between the light on a pillar and that ⟨shadow's⟩ tip.

⟨As for⟩ "**the upright side multiplied by the gnomon**". *Śaṅkuguṇā* is that upright side which has the gnomon for multiplier (a *bahuvrīhi*).

⟨As for⟩ "**Divided by ⟨its⟩ shadow becomes the base**", the base is the height of the light on a pillar. The two shadows also are brought about using their two upright sides. 20

---

[286] An explanation of the mathematical computation described in this rule is given in the supplement for BAB.2.16.

p.94;
line 1

An example :

> **1.The shadows of two same gnomons are observed to be respectively
> ten and sixteen** *aṅgulas* |
> **And the distance between the tips ⟨of the shadows⟩ is seen as thirty.
> Both the upright side and the base should be told.‖**

5   Setting down:

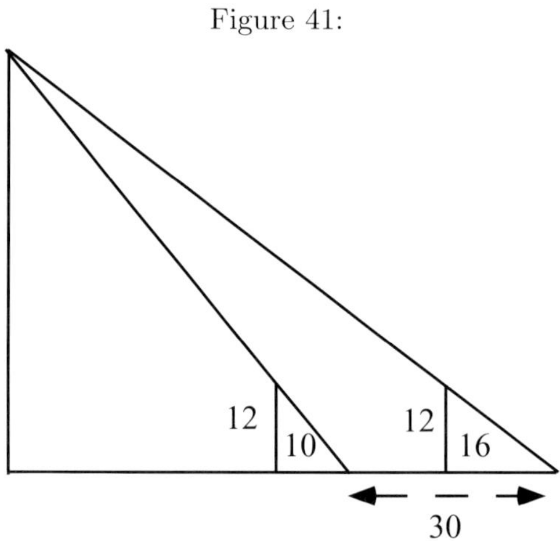

Figure 41:

Procedure: The distance between the tips of the shadows is 30, it is multiplied
by the first shadow, 300; the difference of the ⟨lengths of the⟩ shadows is 6, what
has been obtained with this is the upright side, 50. Precisely this upright side is
multiplied by ⟨the height of⟩ the gnomon (i.e. 12 *aṅgulas*); what has been obtained
is 600, which when divided by the ⟨first gnomon"s⟩ shadow is the base, 60. From
the second shadow, too, the upright side is 80, the base is the same, 60.

An example:

10
> **2. The shadows of two same gnomons (*nara*[287]) are heard of as respec-
> tively five and seven** |
> **The distance between the tips ⟨of the shadows⟩ is seen as eight. Then
> base and the upright side are to be told.‖**

Setting down:

p.95,
line 1

As before what has been obtained is the upright side, 20; and the base 48. With

---

[287]Usually *nara* means "man". However as the height considered in the computation is of 12
aṅgulas or 12 "fingers" this is too small to be actual human beings. We can thus safely assume
that this expression here is used for the gnomons themselves.

Figure 42:

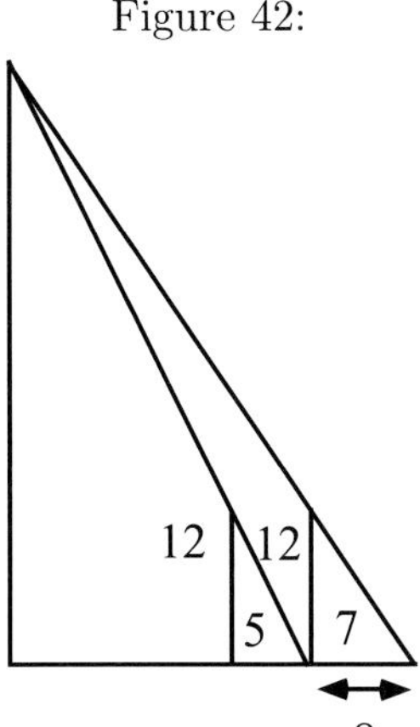

the second shadow too, the upright side is 28, the base is the same, 48.

On an equinoctial (*viṣuvad*) day when the sun (*savitṛ*) is [in the middle] of the vault of the sky, some ⟨people⟩ compute (*ānayāna*) the *yojanas* that make up the distance between the sun and the surface of the earth by means of the *yojanas* that make up the distance between the tips of the ⟨two⟩ shadows which are at ⟨two⟩ spots ⟨located on⟩ an even north-south ⟨line⟩. This is improper. even to resort to mentioning the method with a light and the shadows of two ⟨gnomons⟩ (i.e. Ab.2.16), is not suitable in this case[288].

Why?

Because he has stated: "One should divide the distance between the earth and the sun (...)" [Ab.4.39].

The ⟨diameter of the⟩ earth is a gnomon, the true distance (*karṇa*) in *yojanas* to the sun is the distance between ⟨this⟩ gnomon and the base, the ⟨diameter of the⟩ glorious sun which is the unique light of the whole world, is itself the height of the light. It follows that the computation of the *yojanas* which make the distance between the sun and the surface of the earth is improper, because of the teaching of the *yojanas* that have already been established from the "distance between the earth and the sun"[289].

---

[288] For an explanation of what we understand of the astronomical contents of this paragraph and the following see the supplement for BAB.2.16.

[289] We do not have Bhāskara's commentary to this verse of the *Āryabhaṭīya*. All the manuscripts end at verse 6 of the fourth chapter (on the sphere). The edition of the commentary was completed with the commentary of a later author, Someśvara. This author gives in this commentary to Ab.4.39 a value for the distance between the earth and the sun (p. 278, line 8): 3360 *yojanas*.

But if ⟨it is objected that in the quoted verse⟩ the size of the diameter of the sun is computed (and not therefore the distance between the sun and the surface of the earth) this because ⟨in Ab. 4.39⟩ the sun itself is the "height of the light", this is not so.

Because ⟨the sun⟩ is perceived as having a distance, in *yojanas*, to the earth equal to the true distance of its own orbit, ⟨because it is perceived⟩ as residing in the middle of the sky and as illuminating the worlds, therefore the height of the light cannot ⟨in this case⟩ be the sun itself. Because, that which on its own orbit has the true distance in *yojanas* to the earth is characterized as that which resides in the middle of the sky and illuminates the worlds (i.e. one should distinguish the true sun from the mean sun), therefore the height of the light itself can [not] be the ⟨the diameter of the⟩ sun.

Now ⟨if⟩, the ⟨diameter of the⟩ sun is "the height of the light", ⟨then⟩ the *yojanas* that make up the distance between the centers (*madhya*) of the earth and the sun are the distance to the gnomon having the size of the diameter of the circle (*maṇḍala*) which is the whole surface of the earth, such being the case ⟨the method with two gnomons⟩ is improper because there is no place for a second gnomon. Therefore it has been rightly stated: "even to resort to mentioning the method with a light and the shadows of two ⟨gnomons⟩ (i.e. Ab.2.16), is not suitable".

And the earth has the form of a sphere, this is recited (in Ab.4.6). Therefore, because the circumference ⟨of the earth⟩ has the state of being curved for us who are resting on its back (*pṛṣṭha*), one should assume that the computation (*karman*) of a base and upright with the shadow of a gnomon does [not] occur here, because the configuration of the field having a base, upright side and a hypotenuse (i.e. a right-angle triangle) ⟨is made⟩ with the shadow of a gnomon standing on a spot made flat with water, and it is impossible to make the earth, which is this big, flat.

Now, having asserted ⟨this point⟩, this is stated: On an equinox, in Ujjayinī, when the sun (*uṣṇadīti*) is halfway through the day, the shadow is five *aṅgulas*. The latitude (*akṣa*) obtained with this is twenty-two degrees (*bhāga*) and thirty seven minutes (*lipta*). The distance between the tips of the shadows is ⟨therefore⟩ seventy. With this latitude is obtained the distance in *yojanas* between Laṅkā and Ujjayinī, which amounts to seven-zero (*ambara*)-two (*yama*), 207. Then at the north of Ujjayinī during an equinox, the midday shadow in Sthāneśvara is seven *aṅgulas*. And by means of that, the latitude obtained is thirty degrees and a quarter. With this latitude, the distance between Sthāneśvara and Laṅkā is obtained which amounts to five (*śara*)- seven (*adri*)-two (*yama*), 275.

In this case, the difference of these *yojanas*, which is sixty-eight, is the distance of the two gnomons. Sixty-eight is increased by the difference of shadows, which is two. The distance between the tips of the shadows is ⟨therefore⟩ seventy.

⟨Objection⟩

Here the mathematical computation: "the distance between the tips of the shadows is multiplied by a shadow, etc." (AB.2.16) ⟨can be used⟩, with that computation the upright side has been obtained in *yojanas*. And those ⟨*yojanas*⟩ computed (*nīyamāna*) with the second shadow must be the *yojanas* of the distance between Laṅkā and Sthāneśvara, since at that time the spot standing under the sun is Laṅkā.

p.96; line 1

If the ⟨diameter of the⟩ sun is the base or if the sun's height ⟨is the base⟩, because the upright side is the amount in *yojanas* of the distance between Laṅka and Sthāneśvara, the mathematical computation also is in this case impossible here.

⟨Another proposition⟩

"And in this case, the base is brought about with a certain upright side, which is at first not established; with that ⟨upright side⟩ which is not established, the ⟨already⟩ established base is brought about."

This is improper.

And another ⟨statement⟩: "The computation is made with the shadow in *aṅgulas* of a twelve *aṅgula* gnomon, with the shadow which is known to us by direct perception (*pratyakṣam*), and with the *yojanas* ⟨of the distance⟩."

And this is not suitable.

Now, to say that the shadows of two gnomons measuring twelve *yojanas*, whose shadows are ⟨respectively⟩ five *yojanas* and seven *yojanas*, ⟨should be considered, is not suitable⟩, because the verticality of such a ⟨big⟩ gnomon is impossible to know with a plumb-line, and the uplifting and placing ⟨of the gnomon⟩ are not ⟨possible either⟩. And the shadow is brought about on level ground. In the case of ⟨a distance with⟩ such ⟨a large number of⟩ *yojanas*, there is irregularity due to a depression, an elevated place, a river etc.; therefore its knowledge is impossible. Therefore, the height and the diameter of the sun (*sahasramarīci*) ⟨should be considered to be⟩ exactly as ⟨those⟩ established by tradition. Therefore this net of the method of mathematical calculation (*prakriyā*) should not be stretched over this case.

5

10

[**Relation between the three sides of a right-angle triangle**]

p.96, line 12

In order to compute the hypotenuse, he states:

> **Ab.2.17ab. That which is the square of the base and the square of the upright side is the square of the hypotenuse.|**
>
> yaś caiva bhujāvargaḥ koṭīvargaś ca karṇavargaḥ saḥ|

"That which is the square of the base" and that which is **"the square of the upright side"**, these two squares together (*ekatra*) become **"the square of the hypotenuse"**.

15

An example:

1. The hypotenuses should be indicated in due order for ⟨the trilateral⟩ whose base and upright side are respectively three and four,|
And also for the one whose sizes (*saṅkhyā*) are ⟨respectively⟩ six and eight and for ⟨the one⟩ whose ⟨sides are⟩ twelve and nine.‖

Setting down [290]:

Figure 43:

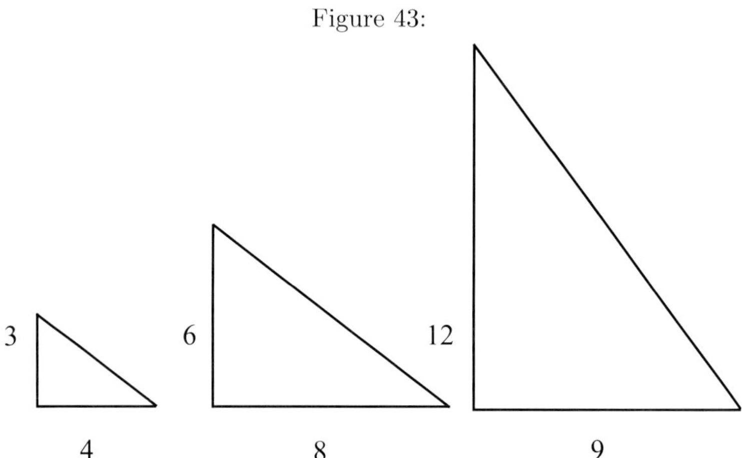

Procedure: These are the base and upright side, 3, 4; their squares are 9, 16; when summed ⟨they become⟩ the square of the hypotenuse 25; its root is the hypotenuse, 5. In this way, the diagonal (*karṇa*) should be considered in a field with an additional half side (*adhyardhā'srikṣetra*) or in a rectangular field (*āyatacaturaśra*). Likewise, in the two remaining fields the hypotenuses obtained are 10, 15.

He states the remaining half of an *ārya* in order to compute (*ānayana*) the chord for a ⟨given⟩ penetration (*avagāha*) in a circular field (*vṛttakṣetra*):

Ab.2.17.cd. In a circle, the product of both arrows, that is the square of the half-chord, certainly, for two bow ⟨fields⟩ ‖

vṛtte śarasaṃvargo "rdhajyāvargaḥ sa khalu dhanuṣoḥ‖

---

[290]Whe have reproduced here the diagrams which have been printed in the edition. The last triangle, however, inverts the base and the upright side. According to the example, the base is 12 and the upright side is 9.

"In a circular" (*vṛtta*) field[291], *śarasaṃvarga* is "**the product of both arrows**" (a *tatpuruṣa*), that is the square of the half-chord. ⟨As for⟩ "**that (. . .), certainly**[292] **for two bow ⟨fields⟩**" becomes the square of the half-chords for precisely those two bow ⟨fields⟩.

10

An example:

> **1. In a field whose diameter is ten, I see two arrows amounting to two and eight|**
> **And in this very ⟨circle⟩ they are also measured as nine and one. The two half-chords should be told in due order.‖**

Setting down:

Figure 44:

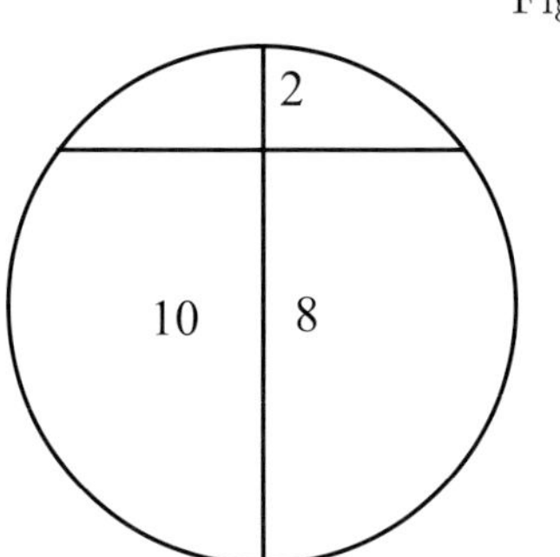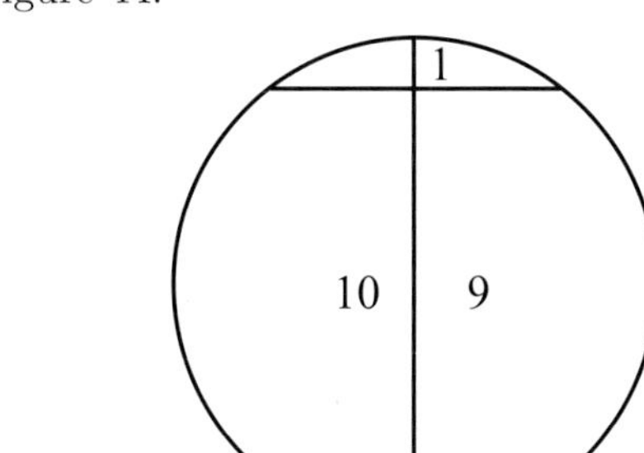

Procedure : There are two arrows 2, 8. Their product is the square of the half-chord, 16. Its root is 4. This is the half-chord. In the second example as well, the half-chord obtained is 3.

p.98, line 1

Just in this case they relate hawk and rat examples. It is as follows: The half-chord is the base. The upright side is in the space between the center of the circle and the half-chord. The hypotenuse which is the root of the sum of their squares, is the semi-diameter of the circle. And this is shown :

5

Setting down:

This half-chord is the height of the hawk's position. The space between the circumference and the half-chord is the rat's roaming ground. The half diameter, which is the hypotenuse, is the hawk's path. The center of the circle is the spot of the rat's slaughter. Here, since the height of the hawk's position is the half-chord, and since

10

---

[291] *Vṛtta* can be understood as an adjective meaning "circular" in which case the commentator supplies to it the word "field". Or it can be considered as a substantiated adjective, and in this case means "circle", as in the verse.

[292] The adverb *khalu* (certainly) is probably an expletive used for the sake of the meter, glossed by Bhāskara with the word *eva* (precisely) another usual expletive. No specific meaning should be given to one or the other expressions.

Figure 45:

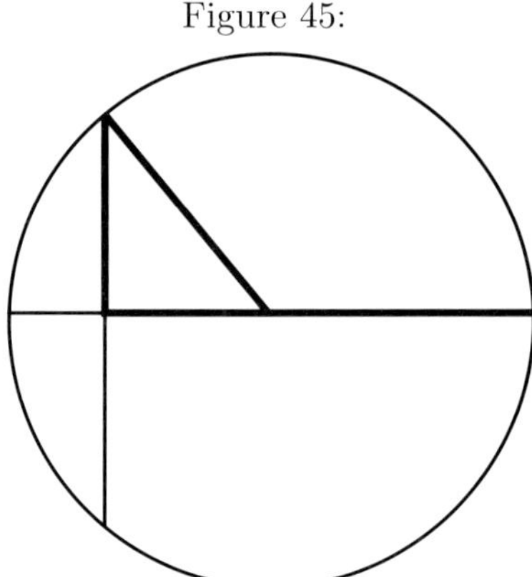

the rat's roaming ground is the arrow, the square of that ⟨half-chord⟩ is divided by
that ⟨arrow⟩. The result is the second arrow. Therefore, when one has performed
that ⟨computation⟩ "⟨it is⟩ increased or decreased ⟨separately⟩ by the difference
⟨and halved⟩" [Ab.2. 24]; the result is the ground up to the rat's residence and the
size of the hypotenuse which is the hawk's [path]. That very second large arrow,
in the quarter of verse on the breaking of a bamboo, is determined as the shape
of a semi-⟨diameter and the side of⟩ a trilateral field (*ardhatryaśrikṣetra*). And
this has been explained. Only the seed of such a computation was instructed ⟨by
Āryabhaṭa⟩.

15   [An example]²⁹³ :

A hawk was resting upon a wall whose height was twelve *hastas*. **The departed
rat [was seen] by that hawk at a distance of twenty** *hastas* **from the foot of the
wall; and the hawk ⟨was seen⟩ by the rat. There, because of his fear, the rat ran
with increasing speed towards his own residence which was in the wall. On the
way he was killed by the hawk moving along the hypotenuse. In this case we wish
to know what is the distance (***antara***) to be attained²⁹⁴ by the rat, and, what is
⟨the distance⟩ crossed by the hawk.**

p.99,
line 1

5   Setting down:

Procedure: The square of the height of the hawk is 144; when that is divided by
the size of the rat's roaming ground, 24, the result is 6. The rat's roaming ground,
when increased by this difference, is 30, and, when decreased is 18. Their halves
are, in due order, the path of the hawk and the distance to the rat's residence, 15,
9.

---

²⁹³This is a non-versified example of the printed edition.
²⁹⁴Reading *prāpya* rather than the *prāpta* of the printed edition.

Figure 46:

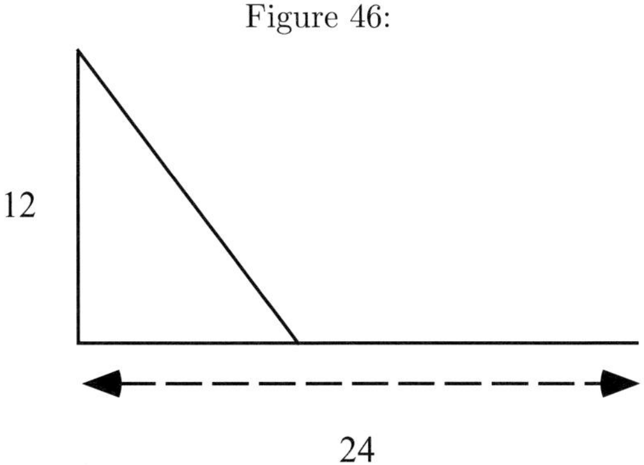

An example:

> **3. A hawk is on a column whose height is eighteen. And there is a rat |**   10
> **He has departed from his residence by eighty one. Because of his fear**
> **of the hawk, ⟨the rat starts running toward the hole⟩ ‖**
> **As he is moving with his residence in his eyes, he is killed on the way**
> **by the cruel ⟨hawk⟩.|**
> **One should then state by what ⟨distance⟩ is the hole reached, and ⟨what**
> **is⟩ the hawk's path.‖**

Setting down: 47

Figure 47:

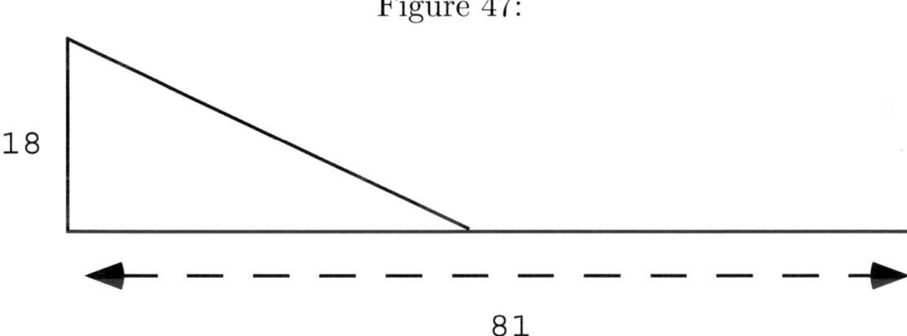

The result is the ground ⟨which should have been⟩ attained by the rat, $1\frac{38}{2}$, ⟨and⟩

the hawk's path $1\frac{42}{2}$.

By means of that very kind ⟨of procedure⟩ an example in the breaking of a bamboo:

> **4. A bamboo with a height of eighteen was broken by the wind,|**
> **It fell, ⟨its tip⟩ having arrived ⟨on the ground⟩ at six from its root, ⟨and**
> **thus⟩ producing a trilateral. Where was it broken? ‖**

p.100, Setting down :
line 1

Figure 48:

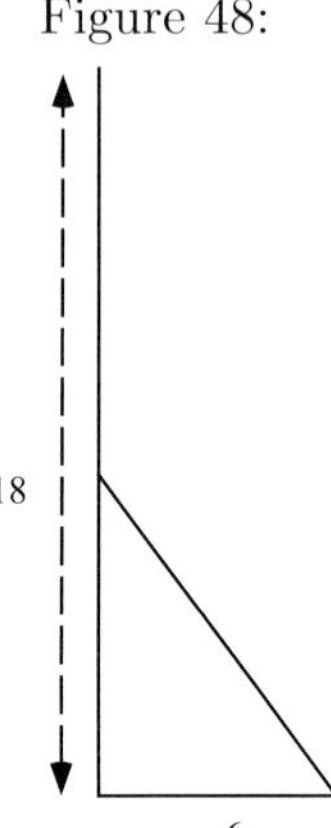

The bamboo is 18, that size which is ⟨the span⟩ of the dropping from the root to the tip is the half-chord, 6. Its square, 36, is divided by that size of the bamboo, 18; the result is 2, as previously, with ⟨the rule⟩ "⟨It is⟩ increased and decreased ⟨separately⟩ by the difference and halved" [Ab.2.24].

5   The two parts of the bamboo are 10, 8.

An example:

> **5. A bamboo of sixteen *hastas* is broken by the wind.|**
> **It fell when ⟨its tip⟩ arrived ⟨on the ground⟩ at eight from its root.**
> **Where was it broken by the one who possesses the wind (*marutvato*,**
> **e.g. the god of the wind)? ⟨This⟩ should be told.‖**

10   Setting down :

The results are the two parts of the bamboo 10, 6.

In examples with lotuses, the size of the seen lotus is one arrow. The extent (*bhūmi*) after which the lotus sinks is the half-chord. In this case, as before, when the square of the half-chord is divided by the ⟨smaller⟩ arrow the largest arrow has been obtained; then with a *saṃkramaṇa* ([Ab.2.24]) using the seen lotus, the size of the water and the size of the lotus ⟨are obtained⟩.

Figure 49:

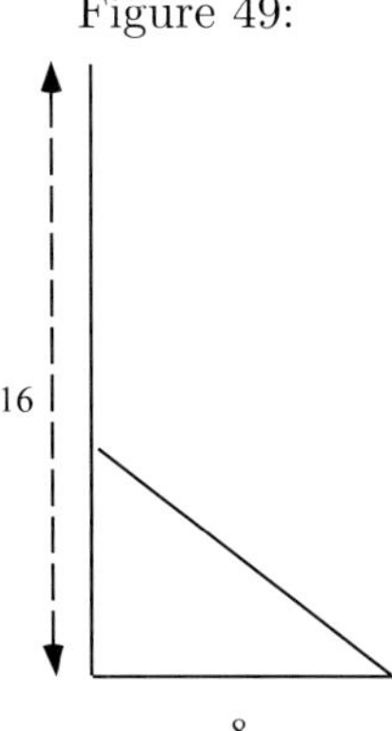

An example :

p.101,
line 1

6. A full bloomed lotus flower of eight *aṅgulas* is seen above the water. Displaced|
By the wind, it sinks in one *hasta*. Quickly, ⟨the sizes of⟩ the lotus and the water should be told.‖

Setting down :

5

Figure 50

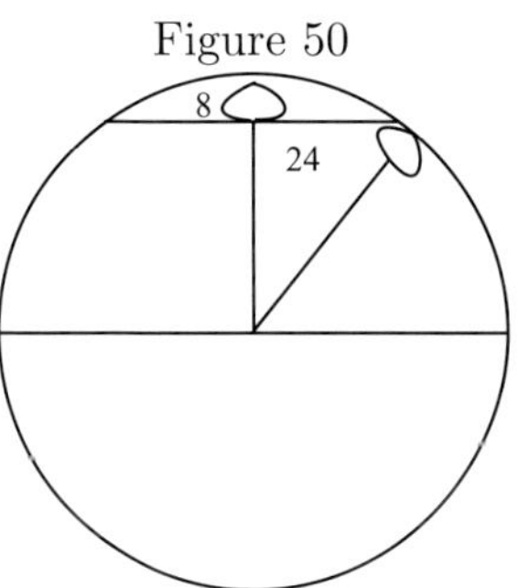

[The size] of the seen lotus is 8, the extent after which it sinks is 24^{295}.

Procedure: The square of the half-chord, which is twenty four, is 576. ⟨It is divided⟩ by eight, which is the seen lotus, the quotient of the division (*bhāgalabdha*) is 72. This is increased by the seen lotus, 80. And decreased, 64. Both are halved; the size of the lotus and of the water are ⟨respectively⟩ 40, 32.

---

[295] One *hasta* is 24 *aṅgulas*.

10   An example:

> **7. A lotus of six *aṅgulas* having moved by two *hastas* from its original
> ⟨spot⟩, sinks.|**
> **In this case I wish to know the ⟨size of the⟩ lotus and the size of the
> water.‖**

p.102,   Setting down :
line 1

Figure 51:

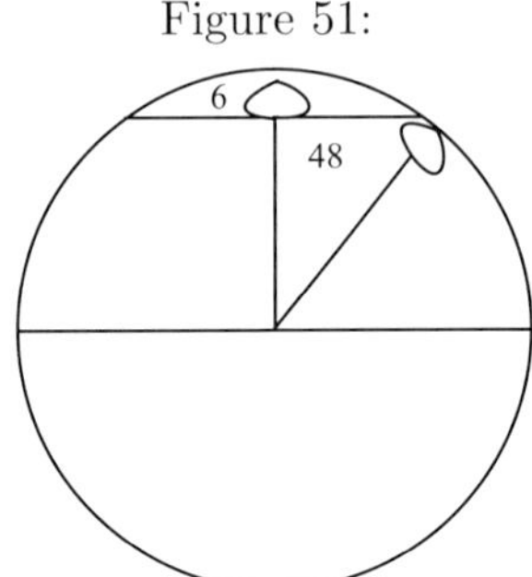

What is seen is 6, the extent ⟨before⟩ sinking, 48. The result, by proceeding as
before, is the size of the lotus 195, the size of the water, 189.

5   In fish and crane examples, exactly in the same way also, the half-chord is one
side (*bāhu*) of a rectangle. The two sides are the greater arrow, what remains is
⟨as⟩ the method for rat and hawk examples.

An example:

> **8. A tank is ⟨measured by⟩ six and twelve. A fish is in its north-east
> ⟨corner⟩.|**
> **In the north-west corner stands a crane. The fish, by fear of that
> ⟨crane⟩, quickly, ‖**

10   **Having cut through the tank diagonally went to the south.|**
> **And was killed by the crane who had moved along the sides. Their
> courses (*yāta*) should be stated.‖**

Setting down:

p.103,   Procedure for the crane and fish : Since the half-chord is a side of the tank, its
line 1   square is 36. Since the ⟨sum of the⟩ two sides is the greater arrow, what results
is 18. The quotient of the division by this is 2. What has been obtained with the
saṃkramaṇa using this ⟨quotient⟩ and eighteen is the size of the paths of the fish
and crane, and the remaining side of the tank 10, 8. When the remaining portion
of the side is subtracted from the side, the remainder is what ⟨was left⟩ for the
fish to reach to the south-west corner.

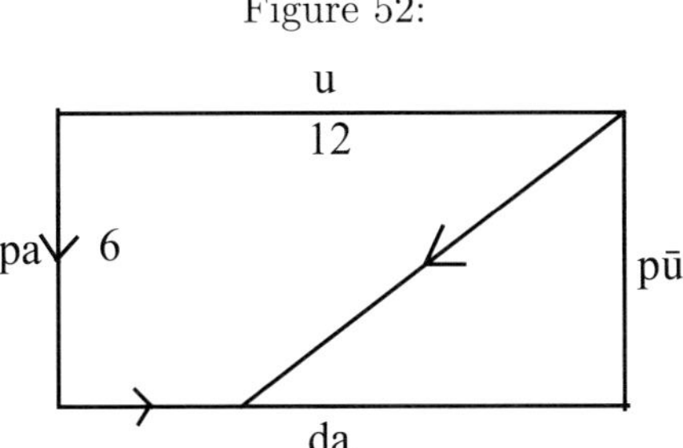

Figure 52:

An example :

9. A tank is ⟨measured by⟩ twelve and ten. Now, a crane stands on the south-east ⟨corner⟩, and there is a fish also|
In the north-east ⟨corner⟩, who went to the western ⟨side⟩ and was killed. To what amounts their two ⟨paths⟩ should be stated.‖

Setting down:

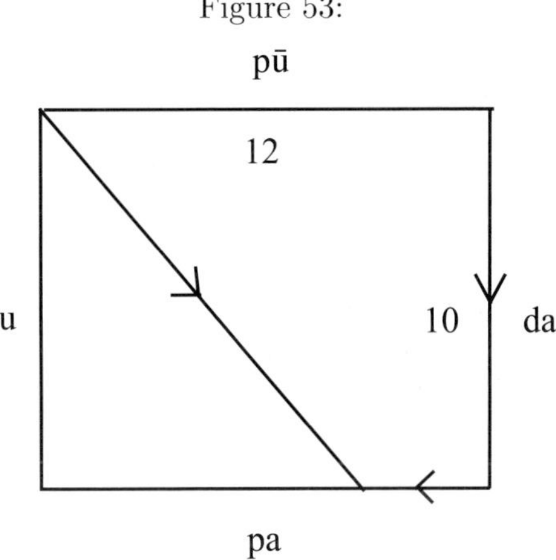

Figure 53:

The result, as before, is the crane's journey from the south-west corner , $\begin{smallmatrix}3\\3\\11\end{smallmatrix}$, the 10 path of the fish which belongs to the western side is $\begin{smallmatrix}8\\8\\11\end{smallmatrix}$. With the procedure which is a seed for reversing unknowns (*vilomabījakaraṇa*) all of this has been brought about.

And the verification within all ⟨geometrical⟩ fields ⟨can be made⟩ with just this ⟨rule⟩ "that which precisely is the square of the base and the square of the upright side is the square of the hypotenuse" [Ab.2.17.ab].

p.103,
line 14   [Knowing the arrow penetrating a circle]

In order to compute the arrow penetrating (*avagāha*) a circle (*vṛtta*) he states:

> **Ab.2.18. One should divide separately the ⟨diameters of⟩ the two circles
> decreased by the *grāsa* and having the *grāsa* for multiplier,|
> The two quotients ⟨of the division⟩ by the sum of ⟨the diameters⟩ de-
> creased by the *grāsa* are the two arrows at the meeting, which are
> ⟨in relation to⟩ one another‖**
>
> grāsone dve vṛtte grāsaguṇe bhājayet pṛthaktvena|
> grāsonayogalabdhau saṃpāta'sarau parasparataḥ‖

p.104,   *Grāsone* is **decreased by the** *grāsa* (an instrumental *tatpuruṣa* in the dual case).
line 1   What are ⟨these which are decreased by the *grāsa*⟩? ⟨They are⟩ *dve vṛtte*, that
is two circles (*maṇḍala*), which are the seized (*grāhya*) and the seizer (*grāhaka*).
Grāsaguṇa is **having the *grāsa* for multiplier** (a *bahuvrīhi* in the dual case).

⟨As for⟩ **"one should divide separately"**, ⟨one should divided⟩ one and the other. By
what? ⟨ he says⟩ **"the two quotients ⟨of the division⟩ by the sum of ⟨the diameters⟩
decreased by the *grāsa*"**. *Grāsonayoga* is the sum (*samāsa*) of those ⟨diameters
of⟩ two circles (*vṛtta*) deprived of the *grāsa* (a dual instrumental *tatpuruṣa*).
Grāsonayogalabdhau is the two quotients ⟨of the division⟩ by the sum of ⟨the
diameters⟩ decreased by the *grāsa* (an instrumental *tatpuruṣa* in the dual case).
Saṃpāta'sarau is **the two arrows at the meeting** (a locative *tatpuruṣa*), it amounts
to the two penetrating arrows. *Parasparataḥ* is ⟨in relation to⟩ one another (*any-*
5   *onya*) .

⟨When a large and a small circle intersect⟩ the arrow of the circle (*maṇḍala*)
having a large diameter (*vyāsa*) is small since the circle is large, and the arrow
of the ⟨circle⟩ having a small diameter is large. The ⟨arc⟩ part (*avayava*) of the
small circle, though small, is perceived to be exceedingly curved, and is not so for
the large ⟨circle⟩. Therefore the two arrows at the meeting are ⟨in relation⟩ to one
another.

An example:

> **1. When eight of the moon (*indu*) whose ⟨diameter⟩ is thirty-two are
> covered by the one ⟨called *Rāhu*[296]⟩ made of darkness whose diam-
> eter is eighty | I wish to know what are the sizes of the arrows of   10
> *Rāhu* and of the moon whose shape is full.‖**

Setting down:

Figure 54:

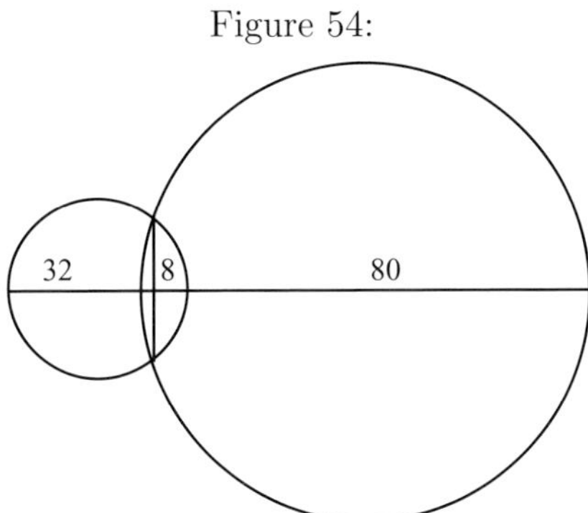

Procedure : The ⟨diameters of⟩ two circles decreased by the *grāsa* are 72, 24; having   p.105,
the *grāsa* for multiplier, 576, 192. The sum of the ⟨diameters⟩ decreased by the   line 1
grāsa is 96. The two quotients ⟨of the division⟩ by that ⟨quantity⟩ are respectively
the moon 6, Rāhu 2, ⟨which are in relation⟩ to one another.

[Value of series]

Now, in order to compute the value of series (*śreḍhīgaṇita*) he states:

> **19. The desired ⟨number of terms⟩, decreased by one, halved, increased
> by the previous ⟨number of terms⟩, having the common difference
> for multiplier, increased by the first term, is the mean ⟨value⟩ | p.105; l.
> ⟨The result⟩, multiplied by the desired, is the value of the desired   5
> ⟨number of terms⟩. Or else, the first and last ⟨added together⟩ mul-
> tiplied by half the number of terms ⟨is the value⟩.‖**

> iṣṭaṃ vyekaṃ dalitaṃ sapūrvam uttaraguṇam samukham[297] madhyam|
> iṣṭaguṇitam iṣṭadhanam tv athādyantaṃ padārdhahatam‖

---

[296] Rāhu is a demon said to swallow the moon during an eclipse.

[297] Here the meter would require that no consonant ends the word, but it seems that Bhāskara
considered that an *ṃ* stood here.

Iṣṭaṃ is wished, *vyekam* is **decreased by one**, *dalita* is **halved** (*ardhita*), *sapūrvam* is **increased by the previous ⟨number of terms⟩**, those [terms] which stand before the desired term (*pada*) are denoted with the word *pūrva*; ⟨the meaning of⟩ *sapūrvam* is that it occurs (i.e. is summed) with the previous ⟨number of terms⟩.

Uttaraguṇaṃ is **having the common difference for multiplier** (a *bahuvrīhi*). ⟨As for⟩ ''**increased by the first term**'' (*samukha*), *mukha* is the first term (*ādi*), and ⟨the meaning of⟩ *samukham* is it occurs (i.e. it is summed) with the first term. ⟨This⟩ is **the mean value** (*madhyadhana*).

Iṣṭaguṇita is **multiplied by the desired ⟨number of terms⟩** (an instrumental *tatpuruṣa*). ⟨This is⟩ the *iṣṭadhana*, that is the value (*dhana*) of the desired number of terms (*gaccha*).

Here, (in this verse), there are many rules (*sūtra*) which stand seperately in the stanza (*muktaka*). Their union (*sambandha*) ⟨is made⟩ according to a ⟨suitable⟩ connection (*saṃyoga*) ⟨of words⟩.

"The desired ⟨number of terms⟩ decreased by one, halved, having the common difference for multiplier, and increased by the first term" is the rule in order to compute the mean value (*madhyadhana*).

"The mean ⟨value⟩ multiplied by the desired ⟨number of terms⟩ is the value of the desired ⟨number of terms⟩" is ⟨the rule⟩ in order to compute the value of the ⟨desired ⟩ number of terms (*gacchadhana*).

"The desired ⟨number of terms⟩ is decreased by one, halved[298], increased by the ⟨number of⟩ previous ⟨terms⟩, having the common difference for multiplier, and increased by the first term", is ⟨the rule⟩ in order to compute the value of the last, the penultimate, etc. ⟨terms⟩ (*antyopāntyadhana*).

"The desired ⟨number of terms⟩ decreased by one, halved, increased by the previous ⟨number of terms⟩, having the common difference for multiplier, increased by the first, and multiplied by the desired ⟨number of terms⟩ is the value (*dhana*) of the desired ⟨number of terms⟩", is a ⟨rule⟩ in order to compute a number of as many terms as desired.

In this way these ⟨rules⟩ are united (*pratibaddha*) within an *āryā*, without a quarter. We will explain these ⟨rules⟩ in due order in nothing but examples.

An example:

> **1. The first term of a series has been seen as two, and the common difference has been told to be three.|**
> **The number of terms is expressed as five. Say the values of the mean and of the whole ⟨number of terms⟩.‖**

---

[298]The word *dalitam* is used in all manuscripts as it is indicated in [Shukla 1976; note 3, p. p.105]; even though Shukla omits it from the main text of his edition.

Setting down: The first term (*ādi*) is 2, the common difference (*uttara*) is 3, the number of terms (*gaccha*) is 5.

Procedure: The desired number of terms is 5, decreased by one, 4, halved, 2, having the common difference for multiplier, 6, increased by the first term, 8; this is the mean value. Just this multiplied by the desired number of terms becomes the whole value (*sarvadhana*), 40.

An example :   p.106,
                                                                               line 1

> **2. They say that the first term of a series is eight, and the common**
> **difference is five.|**
> **The number of terms has been seen as eighteen. The values of the mean**
> **and of the whole ⟨number of terms⟩ should be told‖**

Setting down: The first term is 8, the common difference is 5, the number of terms is 18.

The mean value, obtained as before, is $1\frac{50}{2}$, the whole value 909.   5

An example for the computation of the value of the ultimate, the penultimate etc. ⟨terms⟩:

> **3. The number of terms is twenty-five ⟨for a series⟩ whose first term is**
> **seven and whose common difference is eleven.|**
> **In this case say the value of the ultimate and penultimate ⟨terms⟩**
> **quickly and how much is ⟨the last term⟩ of ⟨the same series with⟩**
> **twenty ⟨terms⟩?‖**

Setting down: the first term is 7, the common difference is 11, the number of terms is 25.

Procedure: the desired ⟨term⟩ is the twenty-fifth term only, and therefore it (the desired number of terms) is one[299] ; one is unity, 1. Precisely, this decreased by one is zero (*śūnya*), 0. Precisely this is "increased by the previous ⟨number of terms⟩" (here 24), it is increased by zero, and therefore twenty-four, 24, "having the common difference for multiplier", 264, is "increased by the first," 271, this is the ultimate value. In this case, when computing the penultimate value, the ⟨number of⟩ previous terms is twenty-three, 23. With these, by proceeding as before, the penultimate value obtained is 260. And for the ⟨series with⟩ twenty ⟨terms⟩, the ⟨number of⟩ previous ⟨terms⟩ is nineteen. With these, by proceeding as before, the value of the twentieth term (*pada*) is 216.

---

[299]Reading *iṣṭaṃ pañcaviṃśati-25-pūraṇam* rather than *iṣṭaṃ pañcaviṃśatiḥ, 25 pūraṇam* of the printed edition.

An example in piles of cubes:

> **There are solid quadrilateral piles (*caturaśraghanāś citayaḥ*) having five, four and nine layers|**
> **⟨The number of⟩ equi-quadrilateral bricks (*samacaturaśreṣṭaka*) broken into unities[302] should be told in due order.‖**

Setting down:

Figure 57:

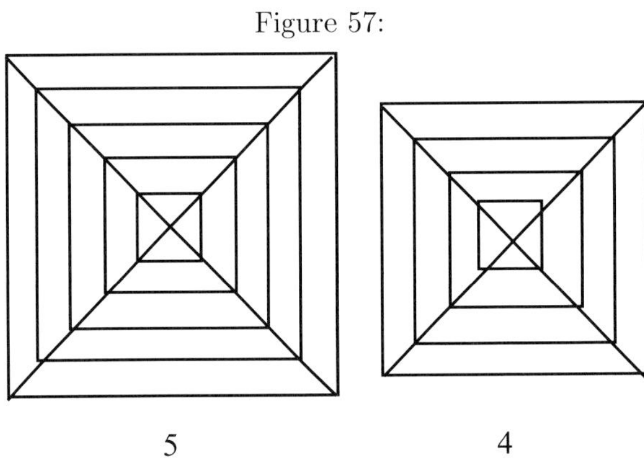

Figure 58:

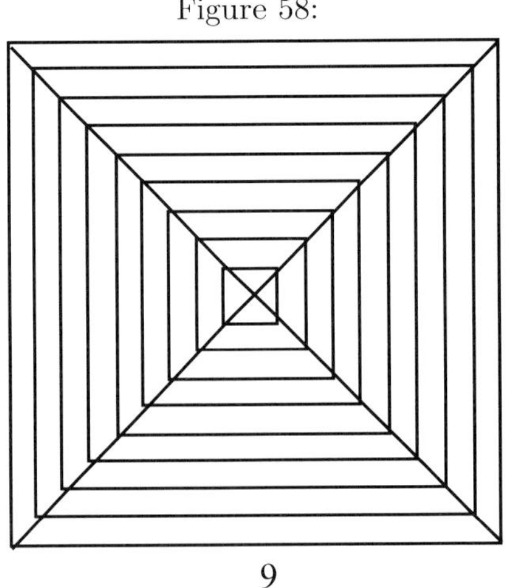

---

[302]Reading *ekāvaghaṭṭitās*, instead of *ekāvaghaṭitās*.

Procedure: The pile (*citi*) is the sum (*saṅkalanā*) [303]. And that is computed with this: "or, the ⟨sum of⟩ the first and the last multiplied by half the number of terms" [Ab.2.19]. In this case the first term is one, 1. The last value is five, 5; the sum (*ekatra*), six, 6, is multiplied by half the number of terms, that is, by the half of five; the pile which is the sum of five is produced, 15; its square is the solid ⟨made⟩ of a pile of cubes. And that is 225.

For the two others also, in due order, what is obtained is 100, 2025.

[Knowing the product of two quantities with another method]

p.112, line 6

In order to compute the product (*saṃvarga*) of two quantities, he states another method (*upāya*):

23. Indeed, one should merely subtract from the square of the sum, the sum of two squares|
That which is its half should be known as the product of two multipliers‖

samparkasya hi vargād viśodhayed eva vargasamparkam|
yat tasya bhavaty ardhaṃ vidyād guṇakārasaṃvargam‖

Samparka is a sum (*samāsa*). Since there is ⟨such a thing as⟩ a sum for two quantities, the sum of only two ⟨quantities⟩ is understood. ⟨The square⟩ of that sum. "**Indeed**" ⟨is used⟩ when filling the verse. *Vargād* is from the square (*kṛti*). *Vi'sodhayed eva* is **one should merely subtract** (*apanayed*). What? He states: "**The sum of two squares**". *Vargasamparka* is the *samparka* or sum (*samāsa*) of two *vargas* or squares (*kṛti*; a genitive *tatpuruṣa*). That sum of two squares should be subtracted from the square of the sum.

⟨As for⟩: "**That which is its half**", that which is half of that, that is, of that which remains from the subtraction; *vidyād* is **should be known** (*anubudadhyād*).

⟨As for⟩: "**The product of two multipliers**". *Guṇakārasaṃvarga* is the product of two multipliers (a genitive *tatpuruṣa*). That product of two multipliers should be known.

10

15

An example:

What will be the product (*ghāta*) of five and four, and of seven and nine|
And of eight and ten?⟨They⟩ should be stated, separately, quickly.‖

Setting down:
$$\begin{array}{ccc} 5 & 7 & 8 \\ 4 & 9 & 10 \end{array}$$

p.113; line 1

---

[303] Reading *citis saṅkalanā* instead of *citisaṅkalanā*

Procedure: The sum of five and four, 9; its square, 81; the square of five, 25; the square of four, 16; together (*ekatra*), 41; one should subtract this sum of the square of five and the square of four from the square of the sum. In this case, the remainder is 40. Its half is the product of four and five; what has been obtained is 20. For the two remaining also, in due order, ⟨what has been obtained is⟩ 63, 80.

7   [Computing the multiplicand and multiplier]

In order to compute two multipliers (*guṇakāra*), he states:

> **24. The square root of the product ⟨of two quantities⟩ with the square of two for multiplier, increased by the square of the difference of the two,|**
>
> **Is increased or decreased by the difference, and halved, ⟨this will produce⟩ the two multipliers of that ⟨product⟩.‖**
>
> dvikṛtiguṇāt saṃvargād dvyantaravargeṇa saṃyutān mūlam|
> antarayuktaṃ hīnaṃ tadguṇakāradvayaṃ dalitam‖

Dvikṛti is the squares of two (a genetive *tatpuruṣa*); *dvikṛtiguṇa* is having the square of two for multiplier (a *bahuvrīhi*). ⟨Of that⟩ having that square of two for multiplier. What ⟨has the square of two for multiplier⟩? He states: "**the product**".

⟨As for⟩ "**Increased by the square of the difference of the two ⟨quantities⟩**"; *dvyantara* is the difference of the two ⟨quantities⟩ (a genetive *tatpuruṣa*); *dvyantaravarga* is the square of the difference of the two (a genetive *tatpuruṣa*). ⟨Increased⟩ by that square of the difference of the two quantities. The square root of the product multiplied by the square of two ⟨and⟩ increased (*miśrata*) by the square of the difference of the two also. That ⟨result⟩ is "**increased by the difference**"; *antarayukta* is increased by the difference (an instrumental *tatpuruṣa*). *Hīnaṃ* is **decreased** (*virahita*). The two multipliers of that ⟨means⟩ the two multipliers of that product. *Dalitam* is **halved** (*ardhita*).

An example:

> **1. In the case where the product is seen as eight, clearly the difference should be two|**
>
> **⟨And then the difference is⟩ seven (*muni*) when ⟨the product is⟩ eighteen. The two multipliers in these two ⟨cases⟩ should be told.‖**

Setting down: The product is 8; the difference 2. The product is 18; the difference is 7.

Procedure: The product is 8; this has the square of two for multiplier, 32; the difference of the two is 2; its square 4; increased by this, 36. Its root is 6; this is increased by that difference, 8; decreased ⟨by that difference⟩, 4. ⟨These two⟩ are halved. In due order, the mutual multipliers are 4 ; 2.

In the second example also two multipliers have been obtained, 9 ; 2.

In this case because there is no difference between the multiplicand (*guṇya*) and the multiplier (*guṇakāra*) both are called multipliers.   25

[Knowing the interest on the capital]   p.114;
                                                                   line 1
In order to compute the interest on the capital (*mūlaphala*), he states:

> **25. The interest on the capital, together with the interest ⟨on the interest⟩, with the time and capital for multiplier, increased by the square of half the capital|**
> **The square root of that, decreased by half the capital and divided by the time, is the interest on one's own capital‖**

> **mūlaphalaṃ saphalaṃ kālamūlaguṇam ardhamūlakṛtiyuktam|**
> **tanmūlaṃ mūlārdhonaṃ kālahṛtaṃ svamūlaphalam‖**

"**The capital**" (*mūla*) is ⟨for example⟩ a hundred, etc.; *phala* is the interest (*vṛddhi*); *mūlaphala* is **the interest on the capital** (a genetive *tatpuruṣa*). "**Together with the interest**" ⟨means⟩ it occurs (i.e. it is summed) with the interest; it amounts to: the interest (*vṛddhi*) on the capital increased by its own interest.

⟨As for:⟩ "**having the time and capital for multiplier**"; *kālamūla* is the time and the capital (a *dvandva*); the interest on the capital is *kālamūlaguṇa*, that is having the time and capital for multiplier (a *bahuvrīhi*).

⟨As for:⟩ "**increased by the square of half the capital**"; [*ardhamūla* is ⟨the same compound as⟩ *mūlārdha*, which is half the capital (a genetive *tatpuruṣa*); *ardhamūlakṛti* is the square of half the capital (a genetive *tatpuruṣa*); ⟨it amounts to:⟩ one fourth of the square of the capital]; because it is the square [of a half], a division is ⟨made⟩ with the square of two, that is, by four; *ardhamūlakṛtiyukta* is increased by the square of half the capital (an instrumental *tatpuruṣa*).

The square-root of what has just been produced is "**the square root of that**". *Mūlārdhona* is "**decreased by half of the capital**", which is a hundred etc. (an instrumental *tatpuruṣa*).

Kālahṛta is **divided by the time** (an instrumental *tatpuruṣa*). *Svamūlaphala* is the interest on one's own capital (a genetive *tatpuruṣa*).

An example:

1. I do not know the ⟨monthly-⟩interest on a hundred. However, the ⟨monthly-⟩interest on a hundred increased by the interest ⟨on the interest⟩ |
Obtained in four months is six. State the interest of a hundred produced within a month‖

$$\begin{array}{cc} 100 & 0 \\ 1 & 4 \\ 0 & 6 \end{array}$$

Setting down: months, 4; interest; 6.

Procedure: The interest on the capital increased by the interest (*phala*) ⟨on the interest⟩, 6; multiplied by the time and the capital, 2400; the square of half the capital, 2500 is increased by that, 4900. Its square root, 70, decreased by half the capital, 20, is divided by the time; the interest on [one's own] capital is what has been produced, 5.

Verification (*pratyayakaraṇa*) with a Rule of Five: "If the monthly interest (*vṛd-dhi*[304]) on a hundred is five, then what is the interest of the interest [of value (*dhana*)-five] on a hundred, in four months?"

$$\begin{array}{cc} 1 & 4 \\ 100 & 5 \\ 5 & 0 \end{array}$$

Setting down:

The result is one. This increased by the ⟨monthly⟩ interest on the capital is six *rūpas*, 6.

An example:

2. The monthly interest on twenty-five is not known. The monthly interest on twenty-five lent at a different place with the same increase (*ardha*), together with the interest (*vṛddhi*) in five months, has been seen as three *rūpas* decreased by one fifth. In this case I would like to know what is the monthly interest on twenty-five, or what is the interest of the interest on twenty-five when lent for five months?

$$\begin{array}{cc} 25 & 0 \\ 1 & 5 \\ 0 & 2 \\ & 4 \\ & 5 \end{array}$$

Setting down:

What has been obtained by proceeding as before is the monthly interest on twenty-five, 2; and the five-month interest of the monthly interest on twenty-five is $\begin{array}{c} 0 \\ 4 \\ 5 \end{array}$ [305].

<hr>

[304] From now on, unless otherwise stated this is the word translated as "interest".

[305] The result could be stated as four fifths. We have discussed this in [Keller 2000; I, 2.2.4.b].

An example:

3. The interest of a hundred in a month is not known. However, the interest of a hundred lent at a different place, together with the interest in five months, has been seen as fifteen *rūpas*. In this case I wish to know what is the monthly interest of a hundred, or what is the interest of the interest of a hundred lent for five months.

$$\begin{matrix} & 100 & 0 \\ \text{Setting down:} & 1 & 5 \\ & 0 & 15 \end{matrix} \quad \text{months, 5; with the interest, 15.}$$

The result[306]: as before, the interest of a hundred is 10; the interest of the interest of a hundred ⟨produced⟩ from an investment on five months is 5.

[The Rule of Three]

p.115,
line 27

In order to explain the Rule of Three (*trairā'sika*) he states an *āryā* and a half:

> **Ab.2.26. Now, when one has multiplied that fruit quantity in the Rule**
> **of Three by the desire quantity|**
> **What has been obtained from that divided by the measure should be**
> **this fruit of the desire‖**
> **Ab.2.27.ab The denominators are respectively multiplied to the multi-**
> **pliers and the divisor.|**

p.116,
line 1

> **trairāśikaphalarāśiṃ tam athecchārāśinā hataṃ kṛtvā|**
> **labdhaṃ pramāṇabhajitaṃ tasmād icchāphalam idaṃ syāt‖**
>
> **chedāḥ parasparahatā bhavanti guṇakārabhāgahārāṇām|**

Trirāśi is three quantities assembled (a *samāhāva dvigu* compound). *Three quantities assembled are the purpose (prayojana) of this computation (gaṇita), and therefore (iti) it is [called] trairāśika. Trairāśikaphalarāśi is* **the fruit quantity in the Rule of Three** (a locative *tatpuruṣa*). That fruit quantity in the Rule of Three.

The word "**now**" (*atha*) is used when finishing a subject of knowledge and explaining the words that follow. Here such a kind of meaning ⟨is understood⟩.

⟨Question⟩

5

What is explained in this case?

This is stated: ⟨The word "now" is ⟩ a speech (*paribhāṣa*). And since this ⟨speech⟩ is different, in each situation (*prativiṣaya*), because of the worldly practise (*lokavyavahāra*), it is explained from ⟨its⟩ use in the world. For if not, a different speech

---

[306]This is an interesting occurence where *labdha* substitutes for the usual *karaṇa* (procedure). Is this a misprint?

⟨would be required⟩ in each situation and the situations are countless. Therefore this ⟨speech⟩ cannot be specified entirely. Accordingly, with the word "now" he (Āryabhaṭa) sets forth the speech as it has been established in the world.

⟨As for:⟩ **"when one has multiplied by the desire quantity"**. That fruit quantity is multiplied by that desire quantity; when one has *hata* that is multiplied (*guṇita*) that by the desire quantity.

"What has been obtained" (*labdha*) is what has been gained (*āpta*). How? He says: **"divided by the measure"**, that is divided by the measure quantity (an instrumental *tatpuruṣa*).

⟨As for⟩ **"from that"**, from such a sort of quantity which has been divided by the measure.

⟨As for⟩ **"the fruit of the desire"**; *icchāphala* is the fruit of the desire (a genetive *tatpuruṣa*); the meaning is: the fruit of the desire quantity.

"This" is said when one has made visible what has been obtained.

⟨Question⟩

Only the Rule of Three is mentioned here by master Āryabhaṭa. How should different proportions (*anupāta*) such as the Rule of Five, etc. be understood?

It is replied: Only the very seed (*bīja*) of proportions has been indicated   by the master, by means of that seed of proportions, the Rule of Five, etc., all indeed, has been established.

⟨Question⟩

Why?

Because the Rule of Five, etc. is a collection of Rules of Three.

⟨Question⟩

How are ⟨these Rules of Three⟩ to be known?

In a Rule of Five, two Rules of Three are collected, in a Rule of Seven, three Rules of Three are collected, in a Rule of Nine, four Rules of Three are collected, and so forth. We will explain ⟨this⟩ only in examples.

⟨Question⟩

When however, the quantities have denominators (*saccheda*), then what should be done?

He says:

> **Ab.2.27.ab. The denominators are respectively multiplied to the multipliers and the divisor.|**

The denominators are *paraspara* multiplied, that is respectively (*anyonya*) multiplied. To what? He therefore says: "**to the multipliers and the divisor**".

Multiplicands (*guṇya*) and multipliers (*guṇakāra*) are multipliers with regard to one another, since ⟨when⟩ the multiplicand is multiplied by the multiplier, and the multiplier also by the multiplicand, there is no difference of results (*phala*) whatsoever. Therefore, what is expressed with the word "multiplier" is the multiplicand and the multiplier.

Guṇakārabhāgahāra is the two multipliers and the divisor.

Therefore, the denominators are respectively multiplied to those multipliers and the divisor; those denominators of the multipliers which are multiplied to the divisor become divisors and the denominators of the divisor multiplied to the multipliers become multipliers[307].

Thought not stated it is clearly understood, since according to their nature (*dharma*) denominators are brought to one or the other ⟨condition⟩. Because the meaning is: the product of divisors is a divisor; the product of multipliers is a multiplier, ⟨the above computation⟩ is understood.

An example:

> **1. I have bought five *palas*[308] of sandalwood for nine *rūpakas*.|**
> **How much sandalwood, then, should be obtained for one *rūpaka*?||**

Here a disposition (*sthāpana*) in due order ⟨should be made⟩. And this has been stated:

> **In order to bring about a Rule of Three the wise should know that in the dispositions|**
> **The two similar (*sadṛ́sa*) ⟨quantities⟩ are at the beginning and the end.**
> **The dissimilar quantity (*asadṛśa*) is in the middle.||**

Setting down: 9   5   1

Procedure: Since five *palas* of sandalwood ⟨have been obtained⟩ with nine *rūpakas*, nine is the measure quantity, five is the fruit quantity. Since "how much ⟨has been obtained⟩ with one *rūpaka*?" ⟨is the question⟩ one is the desire quantity. The fruit quantity multiplied by that desire quantity which is one, 5, is divided by

---

[307]The word *cheda* (denominator) is in the plural form here. This may be understood in two ways: either it refers to the plurality of unities (there is a denominator, and its value is not 1) and therefore only one denominator in fact is considered – this is a current mathematical Sanskrit form. But it can also be understood as a general rule referring to the case where there are several divisors with several denominators. These two interpretations have been further discussed in the supplement for this commentary of verse. The way this half-verse is understood in the Rule of Three is explained in the same supplement.

[308]Concerning units of weight, etc. please refer to the list of Measuring Units in the Glossary.

the measure quantity which is nine, $\frac{5}{9}$. In this case, since parts (*bhāga*) in *palas* are not desired, ⟨one should use:⟩ "a *pala* is four *karṣas*"; ⟨the previous result⟩ multiplied by four is $\frac{20}{9}$. What has been obtained is two *karṣas* and two parts [of nine] *karṣas*. 2 *karṣas* and $\frac{2}{9}$ parts of *karṣas*.

An example:

2. If one *bhāra* of fresh ginger is sold for ten *rūpakas* and one fifth|
Quickly, the price of one hundred and one half *palas*, here, should be
told to me.‖

Setting down:

	2000	10	100
		1	1
		5	2

Disposition in the same category (*savarṇita*):

	2000	51	201
		5	2

"The denominators are respectively multiplied" the denominators of the multipliers go to the divisor. The divisor is multiplied by these two denominators, 5, 2; what is produced is 20000. [The product of one and the other multiplier, 201, 51] is 10251. As before, what is obtained is 10 *vimˊsopakas* and $\frac{251}{1000}$ parts of *vimśopakas*.

p.118, line 1 An example:

3. A *pala* and a half of musk have been obtained with eight and one
third *rūpakas*|
Let a powerful person compute what I should obtain with a *rūpaka* and
one fifth.‖

Setting down: $\begin{bmatrix} 8 & 1 & 1 \\ 1 & 1 & 1 \\ 3 & 2 & 5 \end{bmatrix}$.

With the same category: $\begin{bmatrix} 25 & 3 & 6 \\ 3 & 2 & 5 \end{bmatrix}$.

[By proceeding as before,] what has been obtained by a powerful person is: 13 *māṣakas* of musk, 4 *guñjās* and $\frac{3}{25}$ parts of *guñjās* .

An example:

> **4. A snake of twenty *hastas* moves forward at half an *aṅgula* per *muhūrtta*|**
> **And moves backwards at one fifth ⟨of an *aṅgula* per *muhūrtta*⟩. How**
> **many days ⟨does it take for the snake⟩ to reach the hole?‖**

Setting down: The snake is made of 480 *aṅgulas*, he moves forward at half an *aṅgula* ⟨per *mūhurta*⟩, $\frac{1}{2}$, and moves backward [at one fifth an *aṅgula* ⟨per *mūhurta*⟩, $\frac{1}{5}$].

In this case, since the motion (*gati*) per *muhūrta* of the snake is half an *aṅgula* decreased by one fifth, when one has decreased one fifth from one half, the disposition is: $\frac{3}{10}$, 1 *muhūrta*, the size of the snake in *aṅgulas* is 480.

What has been obtained is $1\frac{53}{3}$ days.   20

In mixed quantities as well, this is the seed of proportions. It is as follows:

An example:

> **5. ⟨A lot of⟩ cattle is said to be ⟨made of⟩ eight tamed, three to be**
> **tamed|**
> **Out of one thousand and one, how many have been tamed and how**
> **many are the others? ‖**   25

Setting down: Eight have been tamed, 8, three are to be tamed, 3, the tamed and un-tamed are one thousand increased by one 1001.   p.119,

In this case, this is the setting down of the Rule of Three: line 1

  Tamed and untamed, 11,   tamed, 8,   the whole collection 1001

In this case the verbal formulation is: "⟨If⟩ with eleven tamed and to be tamed, eight tamed have been obtained, then with one thousand and one how many to be tamed ⟨have been obtained⟩?" The tamed obtained are 728; and therefore the to be tamed are 273.

In a procedure with investments (*prakṣepa*)[309] it is also like this; an example:

> **6. Five connected merchants ⟨invest each⟩ an amount of capital (*mūladhana*)**
> **in ⟨a series⟩ with one for increase and first term (*uttarādi*)|**
> **The ⟨total⟩ profit (*lābha*) is one thousand. Say what should be given to**
> **whom in this case‖**

Setting down: The ⟨respective⟩ amounts ⟨invested for each merchant⟩ are 1, 2, 3, 4, 5. The profit is 1000.

Procedure: With that sum (*prakṣepa*) of amounts, 15, this profit ⟨has been obtained⟩ 1000. The profits obtained, in due order, with one, with two etc. are, [for the first] $66\frac{2}{3}$, for the second $133\frac{1}{3}$, for the third, 200, for the fourth $266\frac{2}{3}$, for the fifth $333\frac{1}{3}$.

An example with fractions (*bhinna*) also:

7. Merchants with ⟨respective⟩ investments (*prakṣepa*) of a half, a third and one eights|
have a profit of seventy minus one. What are their respective ⟨profits⟩?‖

Setting down: $\frac{1}{2}\ \frac{1}{3}\ \frac{1}{8}$, the profit is 69.

In this case, with the method (*nyāya*) for the computation of fractions (*bhinna-gaṇita*): "⟨One and the other quantity⟩ with a denominator has the denominator for multiplier (Ab.2.27cd)" in a same category, what is produced is $\frac{12}{24}\ \frac{8}{24}\ \frac{3}{24}$.

There is no use of the denominators, only the numerators ⟨are taken into account⟩, 12, 8, 3. As before, with the method for investments (*prakṣepanyāya*) , their sum (*ekatra*) is 23. The division of this profit, 69, by that investment, multiplied separately by each one's numerator ⟨is made⟩ with a different Rule of Three ⟨for each merchant⟩, what has been obtained is ⟨the respective⟩ parts ⟨of each merchant⟩ 36, 24, 9.

[The Rule of Five]

An example in the Rule of Five:

8. The interest (*vṛddhi*) of one hundred in a month should be five. Say, how great is the interest|
of twenty invested for six months, if you understand ⟨Ārya⟩bhaṭa's mathematics‖

Setting down: $\begin{matrix} 100 & 20 \\ 1 & 6 \\ 5 & \end{matrix}$

---

[309] For the specific vocabulary used in commercial problems, please refer to the the Glossary.

Procedure: The first Rule of Three is 100, 5, 20. What has been obtained is one *rūpaka*, 1. The second Rule of Three: "If with one month a *rūpaka* ⟨has been obtained⟩, with six ⟨months⟩, how much ⟨will be obtained⟩?" What has been obtained is six *rūpakas.*

This very ⟨series of⟩ computation⟨s⟩ performed simultaneously is a Rule of Five. And here, "of a hundred in a month" ⟨gives⟩ the two measure quantities, [a hundred and one], "five" is the fruit quantity, "what is ⟨the interest obtained⟩ with twenty by means of a six months ⟨loan⟩?" Thus, twenty and six are the desire quantit⟨ies⟩. Here, exactly as before, the desire quantit⟨ies⟩ multiplied by the fruit quantity ⟨are⟩ divided by the two measure quantities[310], the fruit ⟨has been obtained⟩ exactly as earlier. This is nothing but a Rule of Three arranged twice. Furthermore, denominators, as before, go respectively to the divisors and multipliers.

[An example]:

9. Five is the interest of a [two months] investment (*prayukta*) [of one hundred]. What is the interest of twenty-five on a five months investment?

Setting down : $\begin{matrix} 100 & 25 \\ 2 & 5 \\ 5 & \end{matrix}$, what has been obtained is $\begin{matrix} 3 \\ 1 \\ 8 \end{matrix}$.

An example:

10. Four and a half (lit. "the fifth is a half", *ardhapañcaka*) *rūpakas* is the interest of a [three and one half] months investment [of a hundred]. What is the interest of fifty on a ten months investment ?

Setting down: $\begin{matrix} 100 & 50 \\ 3 & 10 \\ 1 & \\ 2 & \\ 4 & \\ 1 & \\ 2 & \end{matrix}$

What has been obtained is six *rūpakas*, 6, and three seventh $\dfrac{3}{7}$.

---

[310]Strangely, the subject of the sentence here ("desire quantity") is in the singular form whereas the "reference quantities" is in the dual case. This is probably due to a corruption of the manuscripts.

p.121, An example :
line 1

**11. A *rūpaka* and one third is the interest produced by twenty and a
half|**

**For a ⟨one⟩ and one fifth months investment. But for seven decreased
by a quarter‖**

**One should tell what is the interest during six months increased by one
tenth|**

**When the arrangement of the denominators (*chedavikalpa*) according
to the *sūtra* in ⟨Ārya⟩bhaṭa's treatise is known. ‖**

$$\begin{array}{cc} 41 & 27 \\ 2 & 4 \\ 6 & 61 \\ 5 & 10 \\ \\ 4 & 0 \\ 3 & \end{array}$$

Disposition in the same category:

What has been obtained is two *rūpakas* [2], 4 [*viṃśopakas*] and $\frac{36}{41}$ [311] parts of

viṃśopakas.

[Rule of Seven]

An example in the Rule of Seven :

**12. When nine *kuḍuvas* of parched and flattened rice are constantly
obtained for an elephant|**

**whose height (*ucchrita*) is seven, circumference (*paridhi*) is thirty and
length (*āyata*) nine‖**

**⟨Then⟩ what should be ⟨given⟩ to an elephant whose height is five,
whose length (*āyāma*) is seven and|**

**circumference is twenty-eight? Then the parched rice obtained should
be told.‖**

$$\begin{array}{cc} 7 & 5 \\ 30 & 28 \\ 9 & 7 \\ 9 & 0 \end{array}$$

Disposition :

What has been obtained is 4 *kuḍuvas* of parched rice, with 2 *setikās* and $\frac{2}{3}$ parts

of *setikās*.

---

[311] The edition reads: $\frac{26}{41}$, probably a misprint.

An example:

**13. When two and a half (lit. "the third is a half", *ardhatṛtīya*) *kuḍuvas*
of beans are obtained for a mighty elephant whose height is four *hastas*,
whose length is six, and circumference (*pariṇāha*) five, then what should
be obtained ⟨for an elephant⟩ whose height is three, length (*āyata*) is
five, and circumference four and a half ?**

Setting down :

$$
\begin{array}{cc}
4 & 3 \\
6 & 5 \\
5 & 9 \\
 & 2 \\
5 & 0 \\
2 & \\
\end{array}
$$

What has been obtained is 1 *kuḍuva*, 1 *setikā*, 2 *mānakas*, half a *mānaka*, $\dfrac{1}{2}$.

In this way ⟨Rules of Three⟩ should be used in the Rule of Nine and the following
ones.

[Reversed Rule of Three]

The reversed Rule of Three (*vyastatrairā́sika*) is just like this also. In this case
the difference is in the inversion (*viparyāsa*) of the multipliers and the divisor. It
is as follows:

[An example:]

**14. Sixteen *palas* of gold are seen when a pala is five sauvarṇikas, then
when ⟨a pala⟩ is four sauvarṇikas, how many ⟨*palas*⟩ of gold are there ?**

Setting down: 5   16   4

In this case, since sixteen *palas* of gold ⟨are obtained⟩ with five *sauvarṇikas*, sixteen
multiplied by five produces the gold ⟨in *sauvarṇikas*⟩. This gold divided by four
produces *palas*, ⟨when⟩ a palas is four *sauvarṇikas*. Therefore 20 *palas* are obtained.

And in his own *Siddhānta* ⟨such a problem requires a Rule of Three:[312]⟩ "when
this *bhujāphala* is obtained in the great circle (*vyāsārdhamaṇḍala*), then how much
⟨is it⟩ in the circle whose semi-diameter (*viṣkambhārdha*) is the hypotenuse produced at that time ?" In this case when the size of the hypotenuse is great, [the
minutes of the *bhujāphala*] become smaller, and when the hypotenuse is small, ⟨the

---

[312]Please refer to the supplement for BAB.2.26-27ab. and the Appendix on astronomy for an
explanation of the computation described here.

minutes of the *bhujāphala*⟩ increase, therefore the semi-diameter is the multiplier [the hypotenuse is the divisor].

An example:

25

15. Eight baskets are seen as measuring fourteen prasṛtikas|
When ⟨each⟩ basket measures eight prasṛtikas, how many baskets there
should then be told.‖

Setting down: 14   8   8 [313]

The result is 14 baskets.

[Fractions in a same category]

He states the latter half of the *āryā* in order to explain an example of fractions (*kalā*) in the same category (*savarṇa*):

Ab.2.27.cd One and the other ⟨quantity⟩ with a denominator has the
denominator for multiplier that is the state of having the same
category‖

chedaguṇaṃ sacchedaṃ parasparaṃ tat savarṇatvam‖[314]

p.123,

line 1   "**⟨Quantity⟩ with a denominator**" (*saccheda*), that is: it occurs with a denominator. What is that ⟨which has a denominator⟩? The integer ⟨part⟩ of the quantity (*rāśirūpa*). In this case, when one has set down (*vinyasya*) the integer ⟨part⟩ of the quantity with a denominator this is said: "**One and the other ⟨quantity⟩ with a denominator has the denominator for multiplier**". *Chedaguṇa* is has the denominator for multiplier (a *bahuvrīhi*), that integer ⟨part⟩ of the quantity with the denominator for multiplier. *Paraspara* is one and the other (*anyonya*), one quantity having a denominator is multiplied by the denominator of another quantity, precisely, one ⟨is multiplied⟩ by the other, such a quantity is set down.

⟨As for⟩ "**That is the state of having the same category**", precisely that accom-

5   plished method (*karman*) is the state of having the same category (*savarṇatva*). According to one's desire the sum (*saṃyoga*) or subtraction (*viśleṣa*) of two ⟨quantities⟩ having the same category ⟨can be made⟩.

---

[313]The setting-down of the printed edition is 8   14   8 ; this does not change the computation as in both cases 14×8 is computed, however the measure quantity and the fruit quantity would here be inverted.

[314]For a mathematical presentation of the contents of this verse, and a discussion please see [Keller 2000; I. 2.2].

An example:

1. One half, one sixth, one twelfth and one fourth are added|
How great is the sum? The value (*dravya*) should be indicated in due
    order‖

Setting down : $\quad \dfrac{1}{2} \quad \dfrac{1}{6} \quad \dfrac{1}{12} \quad \dfrac{1}{4}$ 10

Procedure: For two ⟨quantities⟩ $\dfrac{1}{2} \quad \dfrac{1}{6}$, these two are respectively multiplied by

a denominator, the two quantities with denominators are $\dfrac{6}{12} \quad \dfrac{2}{12}$. Together

(*ekatra*) ⟨it⟩ is $\dfrac{2}{3}$. Disposition with the following third quantity $\dfrac{2}{3} \quad \dfrac{1}{12}$ by

summing ⟨what has been obtained⟩ [is $\dfrac{3}{4}$]. Likewise, with a fourth quantity

$\dfrac{3}{4} \quad \dfrac{1}{4}$, what has been obtained is a *rūpaka*, 1.

An example :   15

2. One half, one sixth with one third are summed. How great is the
    value ?|
And one half, one sixth, one twelfth, one twentieth with one fifth?‖

Setting down: $\quad \dfrac{1}{2} \quad \dfrac{1}{6} \quad \dfrac{1}{3}$

Disposition in the second example : $\dfrac{1}{2} \quad \dfrac{1}{6} \quad \dfrac{1}{12} \quad \dfrac{1}{20} \quad \dfrac{1}{5}$ 20

What has been obtained, by proceeding as before, in both cases is one quantity,
a unit 1, 1.

An example:

How great is the value arithmeticians should count in a half decreased
    by one sixth ,| 25
And in one fifth minus one seventh also or in one third decreased by
    one fourth?‖

Setting down : $\quad \dfrac{1}{2} \quad \dfrac{1°}{6} \quad\quad \dfrac{1}{5} \quad \dfrac{1°}{7} \quad\quad \dfrac{1}{3} \quad \dfrac{1°}{4}$ p.124, line 1

What has been obtained is in due order: $\dfrac{1}{3} \quad \dfrac{2}{35} \quad \dfrac{1}{12}$

p.124,   [Reversed operation]
line 5
He states, in order to teach the reversed procedure (*pratilomakaraṇa*):

> **Ab.2.28 In a reversed ⟨operation⟩, multipliers become divisors and divisors, multipliers |**
> **And an additive ⟨quantity⟩ becomes a subtractive ⟨quantity⟩, a subtractive ⟨quantity⟩ an additive ⟨quantity⟩.‖**
>
> **guṇakārā bhāgaharā bhāgaharās te bhavanti guṇakārāḥ |**
> **yaḥ kṣepaḥ so'pacayo'pacayaḥ kṣepaś ca viparīte ‖**

"**Multipliers become divisors**", those which were multipliers (*guṇakāra*) [become
divisors] in a reversed operation (*pratilomakarman*). "**And divisors, multipliers**",
those which were divisors (*bhāgahāra*) become multipliers. "**And an additive ⟨quantity⟩ becomes a subtractive ⟨quantity⟩**", that which previously was an additive
⟨quantity⟩ (*kṣepa*) becomes the subtractive ⟨quantity⟩ (*apacaya*) in a reversed operation (*vilomakarman*). "**The subtractive ⟨quantity⟩ becomes an additive ⟨quantity⟩**", the subtractive ⟨quantity⟩ becomes an additive ⟨quantity⟩ in the reversed
operation (*viparītakarman*).

Examples for this case have been taught for the most part. And in ⟨our⟩ own
treatise (*tantra*), too, when computing (*ānayana*) the ⟨time in⟩ *ghatikās* from the
Rsine of altitude (*śaṅku*) produced from the Rsine of zenith distance (*chāyā*),
the semi-diameter (*vyāsārdha*) was a divisor and therefore ⟨becomes⟩ a multiplier; the Rsine of the observer's colatitude (*lambaka*) was a multiplier and therefore ⟨becomes⟩ a divisor. In this case, in the northern ⟨hemi-⟩sphere (*gola*), one
had to add the earth sine (*kṣitijyā*), and therefore ⟨it⟩ is subtracted (*apanī-*); in
the southern ⟨hemi-⟩sphere one had to subtract ⟨it⟩ and therefore it is added
(*prakṣip-*). Then, just because it is reversed (*viparīta*), the semi-diameter is a multiplier, the day radius (*svāhorātrārdha*) is a divisor. The sine obtained is made
into ⟨its⟩ arcs (*kāṣṭha*). In the northern ⟨hemi-⟩sphere in arcs the *prāṇas* of the ascensional difference (*cara*) are added, because ⟨they⟩ were subtracted (*viśodhita*);
in the southern ⟨hemi-⟩sphere they are subtracted, because they had the state of
being added, etc...[315]. In this way, everywhere, in ⟨our⟩ own treatise the reversed
operation should be used.

Here is an example in another case :

> **1. Two times ⟨a given quantity⟩, is increased by one, divided by five,
> multiplied by three and again|**
> **Decreased by two, divided by seven; the result is one. How much was
> there before?‖**

---

[315]Please refer to the supplement of this verse for an explanation of the astronomical computation mentioned here.

Setting down : 2 gu[316]; 1 kṣe[317]; 5 hā[318]; 3 gu; 2 ū[319]; 7 hā. And the quotient of the division (*bhāgalabdha*) by seven is one, 1.

This procedure is as follows: the result is one,1; multiplied by seven, what results is 7; increased by two, 9; divided by three, 3; with five for multiplier, 15; decreased by one, 14; halved, what has been obtained is 7.

An example :

> **Three times ⟨a quantity⟩, decreased by one, halved, increased by two and again|**
> **Divided by three, and then, decreased by two, is one. What was there ⟨before⟩?‖**

Setting down: 3 gu; 1 ū; 2 hā; 2 kṣe; 3 hā; 2 ū; the result is 1. What comes forth, as before is 5.

[A particular case of equations with more than one color/category[320]]

He says the computation (*ānayana*) of the sum (*saṅkalita*) concerning the "decreased by a quantity" (*rāśyūna*) method (*krama*):

> **Ab.2.29 The value of the terms decreased by ⟨each⟩ quantity, separately added|**
> **Is divided by the number of terms decreased by one, in this way, that becomes the whole value‖**
>
> **rāśyūnaṃ rāśyūnaṃ gacchadhanaṃ piṇḍitaṃ pṛthaktvena|**
> **vyekena padena hṛtaṃ sarvadhanaṃ tad bhavaty evam‖**

Rāśyūna is **decreased by a quantity** (an instrumental *tatpuruṣa*). With the repetition of "*[rā'syūnaṃ] rāśyūnaṃ*", he shows that the mathematical operation (*gaṇitakarman*) is endless. *Gaccha, pada, paryavasāna* are synonyms.

Gacchadhana is **the value of the terms** (a genitive *tatpuruṣa*), with this "decreased by a quantity"-method (*nyāya*) that which [321] ⟨exists⟩ as far as the terms ⟨exist⟩ is called "the value of the terms".

---

[316] This an abbreviation of *guṇa*, multiplier.

[317] This an abbreviation of *kṣepa*, additive ⟨quantity⟩.

[318] This an abbreviation of *hāra*, divisor.

[319] This an abbreviation of *ūna*, subtractive ⟨quantity⟩.

[320] *anekavarṇasamīkaraṇaviśeṣa*: Shukla seems to be using here a vocabulary that was used by Brahmagupta. I haven't seen it used either by Āryabhaṭa or by Bhāskara.

[321] Reading *yadanena* as in all manuscripts rather than *padadhanam* of the printed edition.

p.125, line 1

5

p.125, line 10

15

"**Added**" (*piṇḍita*) is made in one place (*ekatra kṛta*). With ⟨the expression⟩ "**separately**", he shows an undestroyed disposition for the terms obtained in the "decreased by a quantity" method (*krama*). ⟨There are two things here⟩: the use of an undestroyed disposition and the whole sum ⟨of terms⟩ (*sarvadhana*). When the value of a term (*padadhana*) which has been placed without being destroyed is decreased from the sum of terms, one by one, the values (*dhana*) of ⟨all⟩ terms are produced. If however only the knowledge of the whole value would be ⟨the object of the rule⟩ the measure of the value of the terms would be produced, then ⟨the expression⟩ "separately" would be useless because   without making ⟨them⟩ separately also the whole value would have been established.

"**Decreased by one**" (*Vyekam*) is one has been removed (*vigata*); and by that number of terms it is decreased by one.

⟨Objection⟩

In this case there should be a plural number, "*vyekaiḥ padaiḥ*".

There is no mistake. Having accepted "*pada*" as a class (*jāti*), ⟨as prescribed:⟩ "plural optionally can be used for singular when *jāti* 'class' is to be expressed" ([*Āṣṭādhyāyī*, 1.2.58][322]), the singular number is made. Therefore, the meaning that ⟨is understood from the singular case⟩, "*vyekena padena*", that same meaning is understood from ⟨the expression in the plural form⟩ "*vyekaiḥ padaiḥ*".

Hṛta is **divided** (*bhakta*). ⟨As for⟩ "**The whole value**" (*sarvadhana*). The value in one place, in due order, of all the terms (*pada*), is called "the whole value".

"That becomes the whole value": when this operation (*karman*) is performed in this way, that becomes the whole value.

An example:

> **1. In a forest there are ⟨four⟩ herds of elephants made of ⟨respectively,
>   those⟩ in heat, ⟨those⟩ not in heat, the females and the young;
> They are in this case computed (*gaṇita*) by collecting ⟨them⟩ without
>   one ⟨of the groups⟩: thirty, the square of six (*rasa*)|
> And also ⟨the square⟩ of seven, and just that ⟨last one⟩ increased by
>   one. Let the ⟨whole⟩ assembly (*agra*) of elephants be computed
> And let the computation of each one of the herds be observed accurately.||**

Setting down: 30, 36, 49, 50.

Procedure: These undestroyed sums are together (*ekatra*), 165. The number of terms decreased by one, 3. With that, what has been obtained is the sum ⟨of terms⟩, 55. When one has discarded from this, the first term ⟨of the given series⟩, the assembly of ⟨elephants⟩ in heat is 25; when one has discarded the second, the assembly of ⟨elephants⟩ not in heat is 19; when one has discarded the third, the

---

[322]We have adopted the translation of [Sharma 1990; II p. 129].

number of females is 6; when one has discarded the fourth, the number of young ⟨elephants⟩ is 5.

An example:

> **2. ⟨Seven herds of⟩ elephants, horses, goats, donkeys, camels, mules,**
>   **cows without one ⟨class⟩, in due order, are computed**   15
> **Twenty-eight is the ⟨first sum and it is continuously⟩ decreased by one,**
>   **the last one is again decreased by one. You should definitively tell|**
> **Their whole value and each ⟨class⟩ according to the rule**
> **⟨If⟩ the whole mathematics presented by Āryabhaṭa, was seen ⟨by you⟩**
>   **in the presence of a *guru*.‖**

Setting down: 28, 27, 26, 25, 24, 23, 21.

The whole value obtained is 29; one by one 1, 2, 3, 4, 5, 6, 8.   20

[Equations with one category[323]]   p.127,
                                                                               line 1

In order to show an example of equations (*samakaraṇa*), he states:

> **Ab.2.30 One should divide the difference of coins[324] ⟨belonging⟩ to two**
>   **men by the difference of beads.|**
> **The result is the price of a bead, if what is made into money ⟨for each**
>   **man⟩ is equal.‖**
>
> **gulikāntareṇa vibhajed dvayoḥ puruṣayos tu rūpakaviśeṣam|**
> **labdhaṃ gulikāmūlyaṃ yadyanarthakṛtaṃ bhavati tulyam‖**

With the word "bead" (*gulikā*) an object whose price is unknown is named. *Gu-* 5
likāntara is **the difference of beads** (a genitive *tatpuruṣa*), ⟨one should divide⟩
by that difference of beads; the meaning is: by the difference of ⟨the number of⟩
objects whose price is unknown.

⟨As for⟩ **"one should divide the difference of coins ⟨belonging⟩ to two men"**. With
that "two", he shows that this operation is only for two ⟨men⟩ and not for three
or more. And with that "difference of coins", the wealth (*dhana*) whose amount
(*saṅkhyā*) is known is understood. A coin is, ⟨for example⟩, *dīnāras* etc.

⟨As for⟩ **"the result is the price of a bead"**, that which has been obtained here is
the price of a bead. ⟨As for⟩ **"if what is made into money ⟨for each man⟩ is equal**
(*tulya*)"**, that ⟨wealth belonging to each man⟩, which by means of money, has been

made *tulya*, that is equal (*sadṛśa*).   10

An example:

> **1. The first tradesman has seven horses with perpetual strength and auspicious marks|**
> **And a hundred *dravyas* are seen by me in his hand.‖**
> **Nine horses and the amount of eighty *dravyas* ⟨belonging to⟩ the second ⟨tradesman⟩ are seen.|**
> **The price of one horse, and the equal wealth ⟨of both tradesmen⟩ should be told by ⟨assuming⟩ the same price ⟨for all the horses⟩ ‖**

15

Setting down: 7 100
               9 80

Procedure: The difference of beads, 2; the difference of coins, 20.[325] This divided by the difference of beads is the price in *dravyas* of one horse, ten, 10. With that price, the price of the horses of the first ⟨tradesman⟩ is 70, of the second is 90. With what exists in the hand of each ⟨tradesman⟩ and with this ⟨price of each one's horses⟩, equal wealth exists ⟨in the hands of⟩ both as well, 170.

20

An example:

> **2. Eight *palas* of saffron and ninety *rūpakas* for one offering (*dhana*)|**
> **Twelve *palas* and thirty *rūpakas* for another offering. One should know‖**
> **The saffron that was bought at the same price by two ⟨people⟩ at a certain unknown price per *pala*. |**
> **In this case I wish to know the price and the equal wealth of both.‖**

p.128,
line 1

Setting down: 8 90
               12 30

5

The result, as before, is the price of one *pala* of saffron, 15. The same wealth for both is 210.

Precisely these beads, whose value is unknown, are called ⟨sometimes⟩ *yāvattāvat* (as much as); just ⟨the word⟩ *rūpaka* (coins) ⟨is used⟩ in this case also[326]. examples are told in terms of *yāvattāvat*, too. It is as follows:

[An example:]

10

> **3. Seven *yāvattāvats* and seven *rūpakas* are equal to two *yāvattāvats* [and] twelve *rūpakas*. What is the value of one *yāvattāvat*?**

---

[323]*ekavarṇasamīkaraṇam*: as in BAB.2.29., Shukla seems to be using here a vocabulary that was Brahmagupta's. I have not seen it used either by Āryabhaṭa or by Bhāskara.

[324]Even though a *rūpaka* is a particular coin, since Bhāskara glosses it with *dīnāra* and in examples with *dravya*, he probably understands it here as a coin in general.

[325]As in the verse, the "difference" of coins is expressed by the word *viśeṣa*, whereas the "difference" of beads is expressed by the word *antara*.

[326]Please see the Appendix of BAB.2.30. for an explanation of this sentence.

Setting down: $\begin{matrix} 7 & 7 \\ 2 & 12 \end{matrix}$

Procedure: As before, the difference of beads, or of *yāvattāvats*, subtracted above, is 5. When subtracted below, the difference of coins is 5. The quotient of the division of the difference of coins by the difference of *yāvattāvats* is the value of a *yāvattāvat*, 1. With this value of a *yāvattāvat*, the *yāvattāvats* or beads ⟨of the example⟩ produced are, respectively, 7, 2; which are equal ⟨to each other⟩ when one has added each one's own coins. For the first, 14; for the second, just that, 14.

An example:

4. Nine beads and seven *rūpakas* are equal to three beads|
And thirteen *rūpakas*, then, what is the price of a bead?‖   20

Setting down: $\begin{matrix} 9 & 7 \\ 3 & 13 \end{matrix}$

The result is the price of a bead, 1.

When, on the other hand, *rūpakas* are subtracted, then:

An example :   25

5. Nine beads and the subtractive (*ṛṇaṃ*, lit. debt) twenty four *rūpakas*,
** and, two beads |**
And eighteen *rūpakas* are equal ⟨to each other⟩, [say] what is the price
** of a bead. ‖**

Setting down : $\begin{matrix} 9 & 24° \\ 2 & 18 \end{matrix}$ p.129, line 1

In this case the beads are subtracted above, and the coins which were to be subtracted below, are not to be subtracted.

Then[327]:   5

The debt (*ṛṇa*, prakṛt *bhūṇa*) should be subtracted from the debt, the
** wealth (*dhana*, prakṛt *aṇa* or *dhana* according to manuscripts) from**
** the wealth, ⟨a debt⟩ should not be subtracted from a wealth, ⟨a**
** wealth⟩ not from a debt|**
When it is reversed, just the subtraction ⟨becomes⟩ wealth; nothing is
** concealed, in this case.‖**

In this case the ⟨lower⟩ beads which have to be subtracted above, are subtracted from the ⟨upper⟩ beads, the coins which have to be subtracted below are not be-cause they are debt[328], and because they have to be subtracted, ⟨as the operation⟩

---

[327]This a translation of Shukla 's Sanskrit reading of a very corrupt *prakṛt* verse.
[328]Reading the *śuddhi ṛṇatvāt* of manuscript D.

is reversed (*viparīta*), they are added. When added what results is 42. The quotient of the division by seven with the difference of beads is six, 6. The value (*pramaṇa*) [of the *yāvattāvat* of the first ⟨person⟩] is 9, multiplied by 6, 54. The twenty four *rūpakas*, which are in the state of debt, are subtracted; the remainder is thirty, 30. The two beads of the second are multiplied by six, 12; increased by eighteen, 30. In this way the wealths are equal.

When ⟨dealing with⟩ equations, everywhere, ⟨this procedure⟩ should be used, in this way.

[Knowing the meeting time]

In order to compute the meeting time (*yogakāla*), he states:

**Ab.2.31 When the distance of ⟨two bodies moving in⟩ opposite direc-
tions is divided by the sum of two motions; ⟨or⟩ when the distance
of two ⟨bodies moving in⟩ the same direction ⟨is divided⟩ |
By the difference of two motions, the two ⟨quotients⟩ obtained are the
past or future meeting time of the two.‖**

**bhakte vilomavivare gatiyogenānulomavivare dvau|
gatyantareṇa labdhau dviyogakālāv atītaiṣyau‖**

Bhakta is **divided** (*hṛta*) . ⟨As for⟩ **"when the distance of ⟨two bodies moving in⟩
opposite directions"**. One goes, ⟨while⟩ the other, facing it, comes in the opposite direction, that is "*vilomavivara*", i.e. the distance (*antara*) of a ⟨body moving in⟩ a direct motion (*anulomagati*) and of a ⟨body moving in⟩ retrograde motion (*vilomagati*). Here, it should be understood that the word "direct" (*anuloma*) is dropped but mentioned.

Alternatively, ⟨the expression⟩ "*vilomavivara*" is understood to be of this kind ⟨because⟩ if both ⟨bodies⟩ were in retrograde motions, then it would just be ⟨like⟩ the distance of ⟨two bodies in⟩ direct ⟨motions, and it is not so⟩.

When such a distance of ⟨two bodies moving in⟩ opposite directions is divided. When that distance of ⟨two bodies moving in⟩ opposite directions is divided.

⟨Divided⟩ by what? He says: **"By the sum of two motions"**. *Gatiyoga* is the sum of two motions (a dual genitive *tatpuruṣa*). By that sum of two motions.

⟨As for⟩ **"the distance of two ⟨bodies moving in⟩ the same direction"**. *Anulomavi-vara* is the distance of two ⟨bodies moving in⟩ the same direction (a dual genitive *tatpuruṣa*), meaning the distance of two having a direct motion.

The ⟨expression⟩ "two", refers to[329] the two meetings ⟨computed respectively⟩ for opposite and same ⟨directions⟩.

⟨As for⟩ **"the difference of the two motions"**. *Gatyantara* is the *antara* of two motions (a dual genitive *tatpuruṣa*), i.e. the difference (*viśeṣa*) of two motions. By that difference of two motions.

⟨As for⟩ **"the two ⟨quotients⟩ obtained are the meeting time of the two"**. *Dviyoga* is the meeting of the two (a genitive *tatpuruṣa*). *Dviyogakālau* is the two meeting times of the two (a genitive *tatpuruṣa* in the dual case).

⟨As for⟩ **"the past or future"**. *Atitaiṣya* is *atīta* and *eṣya* (a *dvandva*). *Atīta* is past (*atikrānta*); *eṣya* is future (*bhāvi*). It is as follows : 10

When one planet, standing in the east (*purastāt*), goes in a retrograde ⟨motion⟩ and [the other], situated in the west (*paścād*), goes in an ⟨ordinary⟩ motion, the minutes (*liptās*) of the space (*antarāla*) ⟨separating them⟩ is "the distance of two ⟨bodies moving in⟩ opposite directions"[330].

In this case, since the meeting of one in a direct motion (*anulomacārin*) and of another going in an opposite direction will take place within a short period of time, a division is made by the sum of their daily motions (*bhukti*), because just that much is their daily passing (*āhniko bhogaḥ*).

A Rule of Three is performed, with that ⟨daily passing⟩: If one day has been obtained with that daily passing, then what is ⟨the time obtained⟩ with that distance of ⟨two bodies in⟩ opposite ⟨motions⟩? 15

Days or *ghaṭikās*[331] are obtained. So much time has elapsed when the meeting is passed, and is to come when the meeting will take place.

Here, a method ⟨to compute⟩ the same ⟨longitude⟩ in minutes ⟨for both planets⟩ (i.e. the longitude of the meeting spot, *samaliptās*) by means of the daily motion (*bhukti*) is a Rule of Three: If the true (*sphuṭa*) ⟨daily⟩ motion of a planet has been obtained with sixty *ghaṭikās*, then what is the daily motion ⟨obtained⟩ with the *ghaṭikās* known ⟨in the case of two planets with⟩ opposite ⟨motions⟩?

What has been obtained is summed into the ⟨longitude of⟩ the planet with a direct motion, or subtracted from ⟨the planet with⟩ a retrograde motion. In this way, these two planets, whose ⟨longitudes in⟩ minutes (*liptā*) are the same, are meeting at the desired time[332]. 20

Now, ⟨when a planet with⟩ a retrograde motion stands in the west; ⟨a planet with⟩ a direct ⟨motion stands⟩ in the east, then the result obtained is added into

---

[329]Reading *parāmarśam* rather than *paramaṃśam* of the printed edition.

[330]For an explanation of the different steps of the computations described in the following, please see the supplement for this verse.

[331]Please see the List of Measuring Units in the Glossary.

[332]As stated in the verse quoted below, the meeting time found is an approximation, thus one has to make an extra computation to find an agreeing longitude for the meeting spot.

⟨the longitude of the planet with⟩ a retrograde motion, because the ⟨meeting⟩ is passed; it is to be subtracted from ⟨the longitude of the planet with⟩ a direct motion, precisely because it is passed.

Furthermore, when both ⟨planets⟩ are in a direct motion, then, the division of the distance of ⟨two bodies⟩ with a direct ⟨motion⟩, by the difference of daily motions (*bhuktiviśeṣa*) ⟨is made⟩, because the difference of daily motions is equal to their daily difference of motions (*āhnikaṃ gatyantraram*).

Then, ⟨if⟩ sixty *nāḍīs* (a synonym of *ghaṭikā*) are obtained with that ⟨daily⟩ difference of motions, produced as the difference of daily motions, then what ⟨is the time produced⟩ with the distance of ⟨two bodies with a⟩ direct motion (*anulomavivara*)?

25   *Ghaṭikās* are obtained.

A Rule of Three with those ⟨*ghaṭikās*⟩ together with the true daily motion of the planet: If the exact ⟨daily⟩ motion is obtained with sixty, what ⟨has been obtained⟩ with these *ghaṭikās*?

When the ⟨planet with⟩ a faster (*śīghra*) motion stands westward; the pair is added into the pair, respectively. When the ⟨planet with⟩ a faster motion stands eastwards, that pair is subtracted from the pair. In this way the past or future meeting times of both are produced.

And also when the two have a retrograde motion, then also the operation is just like p.131, that. Precisely this computation has been stated by us in the *Karmanibandha*[333]:
line 1

> **Mbh. 49. If one planet is retrograde and the other direct, divide the difference of their longitudes by the sum of their daily motions.|**
> **Otherwise (i.e., if both of them are either retrograde or direct), divide that by the difference of their daily motions:‖**
> **Mbh. 50. Thus has been obtained the time in terms of days, etc., after or before which the two planets are in conjunction (in longitude).|**
> **The velocity of the planets being different (literally, manifold) (from time to time), the time thus obtained is gross (i.e. approximate).‖**
> **Mbh. 51. One, proficient in astronomical science, should, therefore, apply some method to make the longitudes of the two planets agree to minutes.|**
> **Such a method is possible from the teachings of the preceptor or by day to day practice (of the astronomical science)‖**
> *[Mahābhāskarīya*, 6.49-51][334]

---

[333]The original title of the *Mahābhāskarīya*.
[334]We have adopted Shukla's translation of these verses in [Shukla 1960, p. 201].

For ⟨the meeting⟩ of the sun and the moon also :   10

> **Mbh.4.34. Multiply the unelapsed part of the** *tithi* **or the elapsed part of
> the (next)** *tithi* **by the (true) daily motions of the sun and the Moon
> and divide (each product) by the difference between the (true) daily
> motions (of the sun and the Moon).|**
> **The longitudes of the sun and the Moon increased or diminished (in the
> two cases respectively) by the quotients (thus obtained) should be
> known as the longitudes agreeing to minutes of the sun and Moon-
> the causes of the performances of the world.‖**
> [*Mahābhāskarīya,* 4.64][335]

An example in worldly computations (*laukikagaṇita*) also:

> **1. One ⟨man⟩ goes from Valabhī a** *yojana* **and a half in a day.|**   15
> **And another comes from Harukaccha, ⟨proceeding⟩ a** *yojana* **and a quar-
> ter ⟨a day⟩.‖**
> **And the distance of these two ⟨places⟩ has been observed by travelers
> as being eighteen** *yojanas*|
> **In how much time, mathematician, does the meeting of the two ⟨men⟩
> take place? ⟨This⟩ should be told. ‖**

Setting down: [The ⟨daily⟩ motion] of the one who left Valabhī is $\frac{3}{2}$, [the ⟨daily⟩

motion] of the one who is coming from Harukaccha is $\frac{5}{4}$, their distance for

opposite ⟨motions⟩ is 18.   20

Procedure: The sum of ⟨the daily⟩ motions of the two ⟨men⟩ is $\frac{11}{4}$. The quotient

of the division of, the distance for ⟨two men in⟩ different ⟨motions⟩ by that is 6

days and $\frac{6}{11}$ parts of a day.

An example for a distance of ⟨two bodies moving in⟩ the same ⟨directions⟩:

> **2. A man goes from Valabhī to the Gaṅgā at the daily ⟨motion of⟩ a**
> *yojana* **and a half a day.|**
> **Then another leaves Śivabhāgapura at the daily ⟨motion of a** *yojana*⟩
> **decreased by a third |**   25
> **And their distance is told by wise men to be three times eight** *yojanas.*|
> **In how much time do the two ⟨men⟩, who went along one and the same
> path, meet?‖**

---

[335] We have adopted the translation of *Op-cit*, p. 152.

p.132,
line 1
Setting down : The ⟨daily⟩ motion of the one who left Valabhī is $\frac{3}{2}$, the ⟨daily⟩ motion of the one who left Śivabhāgapura is $\frac{2}{3}$, the distance of ⟨the two men in⟩ direct ⟨motions⟩ is 24.

Procedure: The difference of the ⟨daily⟩ motions of these two ⟨men⟩ is $\frac{5}{6}$, the distance of ⟨the two men in⟩ the same ⟨directions⟩ 24. The quotient of the division of that with the difference of ⟨daily⟩ motions is 28 days and $\frac{4}{5}$ parts of a day.

p.132,
line 5
[Pulverizer]

Now, the pulverizer computation (*kuṭṭākāragaṇita*) is stated. In this case there are two *āryā* rules (*sūtra*):

Ab.2.32 One should divide the divisor of the greater remainder by the divisor of the smaller remainder.|

The mutual division ⟨of the previous divisor⟩ by the remainder ⟨is made continuously. The last remainder⟩ having a clever ⟨thought⟩ for multiplier is added to the difference of the ⟨initial⟩ remainders ⟨and divided by the last divisor⟩.‖

Ab.2.33 The one above is multiplied by the one below, and increased by the last. When ⟨the result of this procedure⟩ is divided by the divisor of the smaller remainder|

The remainder, having the divisor of the greater remainder for multiplier, and increased by the greater remainder is the ⟨quantity that has such⟩ remainders for two divisors‖[336]

10

adhikāgrabhāgahāraṃ chindyād ūnāgrabhāgahāreṇa|
śeṣaparasparabhaktaṃ matiguṇam agrāntare kṣiptam‖
adhaupariguṇitam antyayugūnāgracchedabhājite śeṣam|
adhikāgracchedaguṇam dvicchedāgram adhikāgrayutam‖

⟨As for:⟩ **"one should divide the divisor of the greater remainder"**. *Agra* is a remainder (*śeṣa*). *Adhikāgra* is the greater remainder (a *bahuvrīhi*); *Adhikāgra-bhāgahāra* is that which has a greater remainder and is a divisor (a *karmadhāraya*,

---

[336]For an explanation of the procedure described here please see the supplement for this commentary of verse.

translated as: the divisor of the greater remainder). That divisor of the greater remainder. *Chindyāt*, the meaning is: one should divide (*vibhajed*). By what? He says: "**By the divisor of the smaller remainder. The mutual division by the remainder**". What has been obtained ⟨here, as a quotient⟩ is not used, the operation (*karman*) is carried out with the remainder (*śeṣa*). *Śeṣaparasparabhakta* is the mutual division by the remainder (a double instrumental *tatpuruṣa*) . The meaning is: the division by one and the other (*itaretara*).

⟨As for⟩ "**having a clever ⟨quantity⟩ for multiplier**", the meaning is: having one's own idea (*buddhi*) as a multiplier (*guṇa*).

⟨Question⟩

But how does ⟨a quantity⟩ have one's own idea as a multiplier?

⟨It should answer this question:⟩ Will this quantity (the remainder), multiplied[337] by what ⟨is sought⟩ give an exact division (*śuddhaṃ bhāgaṃ*), when one has added (*prakṣipya*) or subtracted (*viśodhya*) this difference of remainders ⟨to the product⟩?

⟨As for⟩ "**Added to the difference of remainders**"; ⟨it is⟩ added when ⟨the number of placed terms is⟩ even (*sama*), subtracted when uneven (*viṣama*), as is explained by an uninterrupted tradition.

When one has placed in this way the terms (*pada*) obtained by the mutual ⟨division⟩ , the clever ⟨quantity⟩ is placed below, and the last obtained below the clever ⟨thought⟩. ⟨As for⟩ "**the upper is multiplied by the one below**", the quantity above is multiplied by the quantity standing below. ⟨As for⟩ "**increased by the last**", it is increased (*sahita*) by the last quantity, the last one (*paścima*) obtained ⟨as a quotient⟩. In this way, again and again the operation ⟨is repeated⟩ until the computation comes to an end.

⟨As for⟩ "**When ⟨the result of this procedure⟩ is divided by the divisor of the smaller remainder, the remainder**". The remainder, when it is divided by that which is the divisor of the smaller remainder. The remainder of the division of, the quantity produced by means of the previous mathematical operation (*gaṇitakarma*), by the divisor of the smaller remainder is understood.

Having the divisor of the greater remainder for multiplier is multiplied (*abhyasta*) by the divisor of the greater remainder.

Dvicchedāgra is the **⟨quantity that has such⟩ remainders for two divisors** (a genitive dual *tatpuruṣa*); the remainder is a number (*saṅkhyā*). *Adhikāgrayutam* is **increased by the greater remainder** (an instrumental *tatpuruṣa*).

This is what has been stated: When ⟨it is divided⟩ by the divisor of the smaller remainder, the remainder, multiplied by the divisor of the greater remainder and increased by the greater remainder, that is the quantity to be divided by both of these two divisors.

---

[337]Reading *guṇitedam* rather then *guṇitam idam* of the printed edition.

In this way, the pulverizer with remainder (*sāgrakuṭṭākāra*) has been explained. The pulverizer without remainder (*niragrakuṭṭākāra*) will be stated also, afterwards.

5   An example :

> **1. Let a quantity be computed ⟨whose division⟩ by five leaves one unity ⟨as remainder⟩, and|**
> **Whose division by seven ⟨leaves⟩ two unities. In this case, what is the number?‖**

Setting down: $\begin{array}{cc} 1 & 2 \\ 5 & 7 \end{array}$

10   Procedure: The divisor of the greater remainder (*adhikāgraccheda*) is seven, 7. When divided by the divisor of the smaller remainder (*ūnāgraccheda*), 5, the remainder (*śeṣa*) is two above; 2, five below, 5[338]. The quantity (i.e. the remainder) is small[339], therefore just in this case the clever ⟨quantity⟩ is brought about: Will this [above] quantity, multiplied by what ⟨is sought⟩, when one has added the difference of remainders, which is unity, give an exact division by five? The result is the clever ⟨thought⟩, two unities. The quotient of the division is one, 1, the remainder 0. Its placement (*sthāpanā*) is $\begin{array}{c} 2 \\ 1 \end{array}$. Because the third term does not exist ⟨in this line⟩, just this is produced ⟨as the result of the up-going procedure⟩. When divided by the divisor of the smaller remainder, the remainder, 2, is multiplied by seven which is the divisor of the greater remainder, what is produced is

15   14, increased by the greater remainder [2], 16. Just this is the ⟨quantity that has such⟩ remainders for two divisors. Just that quantity when divided by five, has for remainder one, when divided by seven, has for remainder two.

---

[338]This would be a way of stating the fraction obtained from the division of 7 by 5:

$$\frac{7}{5} = 1 + \frac{2}{5}.$$

[339]Concerning the "short-cut" used here please refer to the supplement for this commentary of verse.

An example:

> **2. ⟨A quantity when divided⟩ by twelve has a remainder which is five,
> and furthermore, it is seen by me|**
> **⟨Having⟩ a remainder which is seven, when divided by thirty-one. What
> should one such quantity be?‖**

20

Setting down:
$$\begin{array}{cc} 5 & 7 \\ 12 & 31 \end{array}$$

Procedure: "One should divide the divisor of the greater remainder by the divisor
of the smaller remainder ", the remainder is seven above, twelve below. When the
mutual division ⟨of these two is made⟩ the quotient is one, and again one, the remainder is two above, five below[340]. Here the clever ⟨quantity is computed⟩. There
is an even number of terms, therefore: will this quantity multiplied by what ⟨is
sought⟩ , when one has added the difference of the remainders which is two unities, give an exact division by five ? The result is four unities, which is the clever
⟨quantity⟩. One should place it below the previously obtained ⟨quantity⟩. And
the quotient of the division is two, the result should be placed below. With that
rule (*nyāya*): "The one above multiplied by the one below is increased by the last
remainder", the result is 10. "When divided by the divisor of the smaller remainder", the remainder is just this, which has "the divisor of the greater remainder for
multiplier", what is produced is 310. "Increased by the greater remainder ⟨this⟩ is
the ⟨quantity that has such⟩ remainders for two divisors", and that is this, 317.

p.134,
line 1

5

An example:

> **3. Let ⟨that quantity which divided⟩ by eight has a remainder which is
> five, ⟨which divided⟩ precisely by nine is said to have a remainder
> which is four|**
> **⟨And which divided⟩ by seven has a remainder which is one, be com-
> puted. What should the quantity be?‖**

10

Setting down:
$$\begin{array}{ccc} 5 & 4 & 1 \\ 8 & 9 & 7 \end{array}$$

Procedure:
$$\begin{array}{cc} 5 & 4 \\ 8 & 9 \end{array}$$

---

[340]This would be a way of describing the following set of computations:

$$\frac{31}{12} = 2 + \frac{7}{12},$$
$$\frac{12}{7} = 1 + \frac{5}{7},$$
$$\frac{7}{5} = 1 + \frac{2}{5}.$$

With the pulverizer of these two, [the result] is unity, 1^{341}. And the quantity is thirteen. Here, the product of the ⟨initial⟩ divisors is the divisor ⟨for the remainder, 13⟩, placement: $\dfrac{13 \quad 1}{72 \quad 7}$.

The result as before, for both is the quantity 85. This quantity when divided by eight has for remainder five, ⟨when divided⟩ by nine has for remainder four, ⟨when divided⟩ by seven has for remainder one.

15   An example:

> **4. That quantity which ⟨divided successively⟩ by ⟨the numbers,⟩ starting**
> **with two, and ending with six, leaves a remainder which is one|**
> **And precisely with seven ⟨has an⟩ exact (*'suddha*) ⟨division⟩, say quickly**
> **what it should be, mathematician (*gaṇaka*)!‖**

Setting down: $\dfrac{1 \quad 1 \quad 1 \quad 1 \quad 1 \quad 0}{2 \quad 3 \quad 4 \quad 5 \quad 6 \quad 7}$

In this case, the quantity which has the greater remainder should be chosen according to one's will. The value of the quantity obtained, as before, is 301.

In this way the pulverizer with remainder has been explained.^{342}

p.135,
line 1   [Pulverizer without remainder]

Now, we will explain these two same rules (*sūtra*) as meaning the pulverizer without remainder (*niragrakuṭṭaka*)^{343}:

---

^{341}In this case both the clever quantity and the associated quotient obtained may have been one.

^{342}Example of manuscript E note 7, p. 134:

> **⟨A quantity⟩, respectively divided by ⟨the numbers⟩ starting with two and ending**
> **with nine will have for remainder one.|**
> **You, say quickly what is that numerical quantity respectful one,whose *tilakas* (the**
> ** *tilaka,* is an auspious mark bared on the forehead) are the heavenly bodies!‖**

^{343}The following would be Bhāskara's interpretation of the verses in this case, omitting the last two quarters of verse 33:

> **32. One should reduce the divisor which is a large number ⟨and the dividend⟩ by**
> **a divisor which is a small number.|**
> **The mutual division of the remainders ⟨is made continuously. The last remainder⟩**
> **having a clever ⟨quantity⟩ for multiplier and added in the inside of a number**
> **⟨is divided by the last divisor⟩. ‖**
> **33ab. The one above is multiplied by the one below, and increased by the last.**
> **When ⟨the remaining upper quantity⟩ is divided by the divisor which is a small**
> **number, the remainder is ⟨the pulverizer. When the lower one remaining is**
> **divided by the dividend the quotient is produced.⟩ |**

⟨As for⟩ **"One should divide the divisor which is a large number"**, the meaning is: one should reduce (*apavartayet*). By what? He says: **"By a divisor which is a small number"**. *Agra* is a number (*saṅkhyā*); *ūnāgra* is that which is small and an *agra* (a *karmadhāraya*, translated as: a small number), *ūnāgrabhāgahāra* is the divisor which is a small number (a *karmadhāraya*); by that divisor which is a small number. The meaning is: one should reduce ⟨by that divisor which is a small number⟩. As for example, twenty one is reduced by seven; the dividend also should   5
be reduced by that very ⟨number⟩ with which the divisor is reduced.

⟨Question⟩

How is this ⟨operation⟩, "the dividend also should be reduced by that very ⟨number⟩ with which the divisor is reduced", known?

From an uninterrupted tradition.

Or else, this rule (*nyāya*) ⟨should be understood:⟩ "for the reduced divisor, the reduced dividend must be produced", as for example, one twentyone-th reduced by seven is one third.

Or else, the reduction of the dividend also is mentioned by the teacher who stated the reduction of the divisor.

Why?   10

Because the divisor and the dividend move together (*sahacāritvāt*). As when it is stated: "Let the dishes[344] be washed", the glasses are washed also.

With the words (*grantha*) "The divisor which is a large number", etc., he (Āryabhaṭa) explains this: the pulverizer of the reduced divisor and dividend ⟨is performed⟩.

"The mutual division ⟨of the previous⟩ remainders" is the mutual division of the divisor and the dividend.

"Having a clever ⟨quantity⟩ for multiplier", this is the same as before. ⟨As for:⟩ **" thrown in the inside of a number"**. *Agra* is a number. *Agrāntara* is the inside of an *agra* (a genitive *tatpuruṣa*). The meaning is: the inside of a number (*saṅkhyā*).   15
And that ⟨last remainder⟩ multiplied by a desire ⟨quantity⟩ (*icchā*), when one has added (*prakṣipya*[345]) or subtracted (*apanīya*) this inside number (*saṅkhyāntara*) will give an exact division of that quantity.

"The one above is multiplied by the one below and increased by the last", all of this is the same as before. ⟨As for: ⟩ **"when ⟨the result of this procedure⟩ is divided by the divisor which is a small number, the remainder"**, the meaning is: the remainder of the reduced divisor; "is the pulverizer" is the remaining part of the sentence. The upper [quantity should be made to be] divided by the divisor; the lower quantity should be divided by the dividend quantity. In ⟨a treatise on⟩ mathematics, this has been stated also, in this way, with the words beginning with:

---

[344]Reading "*sthālānī*" as in Ms D, rather than "*sthalānī*" as in the printed edition.

[345]From now on, unless otherwise mentioned this is the verbal root used for this operation.

> **And when the upper is divided by the divisor, then ⟨this⟩ should be
> the quantity|**

The two remainders are the pulverizer (*kuṭṭakāra*) and the quotient (*bhāgālabdha,*
lit. what has been obtained from the division).

"Multiplied by the divisor of the first remainder", etc this is not [used] in pulver-
izers without remainders.

It is as follows, an example:

> **5. Eight multiplied by what ⟨is sought⟩ increased by six, divided by
> thirteen|**
> **should give an exact division. What is the multiplier? and what has the
> quotient (*āpta*)? ‖**

Setting down: $\begin{smallmatrix} 8 & 6 \\ 13 & \end{smallmatrix}$

The dividend is eight, the divisor is thirteen, the inside number is six.

Procedure: The divisor and dividend quantities are reduced by unity $\begin{smallmatrix} 8 \\ 13 \end{smallmatrix}$. "The

mutual division of the remainders ⟨is made continuously⟩", what results is $\begin{smallmatrix} 1 \\ 1 \\ 1 \\ 1 \end{smallmatrix}$.

The remainder of the mutual division is $\frac{1}{2}$. "Having a clever ⟨quantity⟩ for
multiplier and added to the inside number"; will this one quantity multiplied by
what ⟨is sought⟩, when one has added six units ⟨to it⟩, give an exact division by
two? The clever ⟨quantity⟩ is two, 2; ⟨one⟩ is multiplied by the clever ⟨quantity⟩,
what results is $\frac{2}{2}$. This is increased (*yuta*) by six units $\frac{8}{2}$. The quotient is four
units, 4. All of these are, in due order,

$$\begin{matrix} 1 \\ 1 \\ 1 \\ 1 \\ 2 \\ 4 \end{matrix} \ .$$

"The one above is multiplied by the one below and increased by the last", what
results is $\begin{smallmatrix} 22 \\ 14 \end{smallmatrix}$. "When ⟨the result of this procedure⟩ is divided by the divisor
which is a small number, the remainder", the remainders of the divisions ⟨of the
upper and lower quantities respectively⟩ by the divisor and dividend which are

small numbers [= is reduced] are placed (*sthāpita*) $\frac{9}{6}$. This is the pulverizer and   15
the quotient.

An example:

> **6. Eleven, multiplied by what ⟨is sought⟩, decreased by three, these,**
> **divided by twenty-three |**
> **Should give an exact division; tell me the quotient and the multiplier**
> **(*guṇaka*).‖**

Setting down : $\frac{11}{23}$ The quotient ⟨is obtained⟩ from ⟨this eleven⟩ decreased [by   20
three]. By proceeding just as before, the pulverizer and the quotient are $\frac{17}{8}$.

[Planet's pulverizer, method for the residue of revolutions]

p.136,
line 16

Now, in this case, the pulverizer is applied in the mathematics of planets (*graha-gaṇita*, or planetary computations): Will the revolutions of the sun ⟨in a yuga⟩ (*ravibhagaṇa*), multiplied by what ⟨is sought⟩, when one has removed the residue of revolutions (*maṇḍalaśeṣa*), give an exact division by ⟨the number of⟩ terrestrial (civil) days ⟨in a *yuga*⟩ (*bhūdivasa*)?[346] The revolutions of the sun ⟨in a yuga⟩[347] and the ⟨the number of⟩ terrestrial days ⟨in a *yuga*⟩ are set down: $\frac{4320000}{1577917500}$.

p.137,
line 1

These two should be divided by one another in order ⟨to obtain⟩ the divisor which is a smaller number (*ūnāgraccheda*). What remains is the divisor which is a smaller number, seventy five hundred, 7500.[348] Both are reduced by this $\frac{576}{210389}$. These two remainders have been divided by the divisor which is a smaller number. The pulverizer of the illustrious Bhāskara is to be brought forth for both of these.

An example:

> **7. The mean ⟨position⟩ of the sun (*ravi*) produced at the time of sunrise|**   5
> **Is seen by me half way in the Sagittarius (*dhanu*) part of Leo (*mṛgapati*)[349]‖**
> **Let the number of days (*dinagaṇa*) ⟨passed since the beginning of the**
> **Kaliyuga⟩ established in ⟨Ārya⟩bhaṭa's treatise,|**

---

[346] Please see the section on the Astronomical applications of the pulverizer in the supplement for this commentary on verse, for an explanation of the computation carried here and in the following pages.

[347] This value is given by Āryabhaṭa in Ab.1.3.

[348] This is the Greatest Common Divisor of the above numbers. The above sentence would indicate that it was obtained by what is commonly called the "Euclidean Algorithm".

[349] That is the ninth part of the nine divisions of the sign Leo, as it is explained by Shukla in his resolution of the problem, p. 312 of his edition of Bhāskara's commentary. I have not found any description of these subdivisions in Bhāskara's works.

And its (the sun's) passed revolutions (*bhagaṇa*) ⟨established in the⟩ time of the *Kali*⟨*yuga*⟩, be computed‖

Setting down[350]: [The mean sun] 4 | 28 | 20 |

"Multipliers ⟨become⟩ divisors" [Ab.2.28] ⟨is used for⟩ the computation of the residue of revolutions[351] (*maṇḍala′seṣa*).

It is as follows: When one has put the ⟨mean⟩ sun (*savitṛ*) into minutes, the result is 8900. This is multiplied by 210389 (the abraded number of civil days) and divided by zero (*kha*)-zero-the cube of six [21600]; the quotient is the residue of revolutions 86688[352]. When one has selected this very residue of revolutions as the interior of a number, a pulverizer is performed.

[Setting down:] $\begin{matrix} 576 \\ 210389 \end{matrix}$ the interior of a number 86688

The result with the pulverizer rule (*nyāya*) is $\dfrac{8201068565}{22452768}$.

When one has divided by the divisor which is a small number, the two remainders of the divisions are:[$\dfrac{105345}{288}$].

[288], this is ⟨the number of revolutions of the sun⟩ passed in the *Kali*⟨*yuga*⟩, and the number of days ⟨passed in the *Kaliyuga*⟩ (*ahargaṇa*) is 105345.

Or else, when one has performed a pulverizer with one unit as subtractive (*apacaya*), the number of days ⟨since the beginning of the *Kaliyuga*⟩ and ⟨the number of solar⟩ revolutions ⟨since the beginning of the *Kaliyuga*⟩ are computed. It is as follows: With one as subtractive the pulverizer and the quotient are: $\dfrac{94602}{259}$.

And a Rule of Three with this and ⟨the previous⟩ residue of revolutions: If this pulverizer ⟨is obtained⟩ with one as subtractive, how much is it with the residue of revolutions as subtractive?

Setting down: 1, 94602, 86688.

In this case, the result is divided[353] by the reduced (*nirapavartita*[354]) days (i.e. the abraded civil days in a *yuga*), the remainder is exactly the previously written ⟨number of days elapsed since the beginning of the *Kaliyuga*⟩.

---

[350]These values correspond to the following: The ninth part of the *Mṛgapati* sign extends from 146°40′ to 150°, so the middle longitude of this sign is 148°20′, or 4 × 30° (4" signs") +28°20′.

[351]There does not seem to be such a rule to invert in the *Āryabhaṭīya*. This may refer to a rule given by Bhāskara elsewhere in the commentary. A rule for the residue of revolutions is given in the *Mahabhaskarīya*. Please see the section on the Astronomical applications of the pulverizer in the supplement for this commentary on verse.

[352]This is an approximation of the result of division which is: 86688.060185...

[353]Reading: *vibhaktaṃ śeṣam* rather than *vibhaktaśeṣam*.

[354]From now on, unless stated, this is the word used to indicate that the value considered is abraded.

When computing ⟨the number of solar⟩ revolutions ⟨a Rule of Three is used also: If⟩ this quotient ⟨has been obtained⟩ with one unit as subtractive, how much is it with the residues of revolutions ⟨as subtractive⟩?

Setting down: 1, 259, 86688.

In this case, the result is divided by the reduced ⟨number of solar⟩ revolutions ⟨in a *yuga*⟩; the remainder is precisely the previously written ⟨number of solar⟩ revolutions ⟨performed since the beginning of the *Kaliyuga*⟩.

[Method for ⟨the remaining part of⟩ a revolution to be accomplished]

5

Or else when one has selected the two quantities as the reduced divisor and dividend, and [the ⟨remaining part of⟩ a revolution to be accomplished], a pulverizer is computed. It is as follows:

An example:

8. It is said that a hundred minutes (*kalā*) of the eighth sign (*bhavana*)
   are to be crossed by the sun;
Say quickly, having considered ⟨the problem carefully⟩, if you know the
   ***Āśmaka's* (i.e Āryabhaṭa's) mathematics|**
all the years of the *Kali*⟨*yuga*⟩ passed up to this day, clever one,
And let that group (i.e number) of days of the *Kali*⟨*yuga*⟩ which has
   passed be told to me with clarity.‖

10

Setting down: ⟨what is⟩ to be crossed by the sun[355] $\begin{smallmatrix} 7 \\ 1 \\ 40 \end{smallmatrix}$.

⟨Using⟩ "multipliers ⟨become⟩ divisors", with that ⟨part of⟩ a revolution to be accomplished (*maṇḍalagantavyam*), the ⟨part of⟩ a revolution to be accomplished ⟨to be used in the pulverizer procedure⟩ is 123701 [356]. And with that additive remainder, as before, the number of days (*ahargaṇa*) and ⟨the number of revolutions⟩ elapsed in the *Kali*⟨*yuga*⟩ are[357] $\dfrac{105345}{288}$.

15

---

[355] A 100 minutes of the eighth sign is equal to 7 signs, one degree and 40 minutes.

[356] As in the previous case, this is an approximation of the quotient: $\dfrac{12700 \times 210389}{21600}$.

[357] The pulverizer procedure is applied to the following problem, as we algebrise it:

$$y = \frac{576x + 123701}{210389}.$$

The result set down would correspond to x and $y - 1$, as what is considered here is the additional part that would make the number of revolutions integral. The longitude of the sun is the same as in the previous example so that, the results obtained are the same.

The pulverizer and the quotient with one unit as additive[358] are $\dfrac{115787}{317}$. As
before, with that also, ⟨a Rule of Three:⟩ if this pulverizer or quotient of the
division ⟨has been obtained⟩ with one unit as additive, what is the pulverizer or
quotient with ⟨part of⟩ a revolution to be accomplished as additive? When one
has divided the result by the reduced divisor and dividend, the number of days
⟨elapsed since the beginning of the *Kaliyuga*⟩ and the quotient ⟨are obtained⟩ .

In this case, because ⟨part of⟩ a revolution to be accomplished is added, one
becomes additive ⟨to the actual number of revolutions⟩. Therefore one revolution
is subtracted ⟨from the result⟩.

In this way the revolution's pulverizer is explained.

[Sign-pulverizer]

Now, however, the sign-pulverizer is stated. It is as follows:

An example :

> **9. A storm has taken away the signs together with the revolutions
> of the lord of the day (*bhartur divasasya*, i.e. the sun) obtained
> according to the day quantity (i.e. the number of days elapsed since
> the beginning of the *Kaliyuga*)|**
> **What remains   is three times seven and nine-five ⟨respectively⟩ degrees
> and *liptās* ⟨of the sun's longitude⟩;**
> **Say where the sun has gone (i.e. its longitude) together with the day
> quantity.∥**

Setting down: $\begin{array}{c} 0 \\ 0 \\ 21 \\ 59 \end{array}$.

Procedure: "Multipliers ⟨become⟩ divisors", the residue of signs (*rāśiśeṣa*) obtained
is, 154168[359]. Since the reduced (*apavartita*) revolutions of the sun with twelve for
multiplier are the signs[360], the placement (*sthāpanā*) is: $\dfrac{6912}{210389}$; the residue of
signs is 154168.

---

[358]If one is a subtractive we know that the solutions are $\dfrac{94602}{259}$. From these we can deduce
the values for one as an additive: $\begin{array}{c} 210389 - 94602 = 115787 \\ 576 - 259 = 317 \end{array}$.

[359]154168 is an approximation of $\frac{1319 \times 210389}{1800}$. 1800 is the number of minutes in a sign.

[360]$576 \times 12 = 6912$ is the reduced number of signs the sun has passed since the beginning of
the *Kaliyuga*.

The result by undertaking a pulverizer is the number of days ⟨elapsed since the beginning of the *Kaliyuga*⟩ and the quotient : $\frac{176564}{5800}$.

When one has divided the quotient by twelve, the quotient is the passed revolutions (*bhuktabhagaṇa*). The remainder is the signs. These are 483, [4]. The number of days ⟨elapsed since the beginning of the *Kaliyuga*⟩ is [176564. Or else] until it pleases the inquirer (*pṛcchaka*), ⟨the values should be increased by multiples of the constants⟩.

When one has performed also a pulverizer with one unit as subtractive, what has been obtained is: $\frac{113078}{3715}$.

With a Rule of Three[361] with just both of these, the number of days ⟨elapsed since the beginning of the *Kaliyuga*⟩ and the quotient are: $\frac{176564}{5800}$.

What remains is the same.

[Method to secure ⟨the result⟩ with another kind ⟨of procedure⟩]

Others however, when one has performed a pulverizer of twelve and of the terrestrial days with one as subtractive, perform a Rule of Three.

Because twelve is the multiplier of the residue of revolutions. In this case the signs passed and the residue of revolutions are obtained. It is as follows :

Setting down[362]: $\frac{12}{210389}$.

In this case the result is the pulverizer and the quotient of the division: $\frac{122727}{7}$.

What remains is not stated because it has been ⟨previously⟩ told.[363]

---

[361]Multiplying each result by the residue of signs and considering the remainders of a division respectively by the number of terrestrial days in a *yuga* and by the number of signs crossed by the sun in a *yuga*, the following values are found.

[362]The pulverizer is to be applied to:

$$y = \frac{12x - 1}{210389}.$$

[363]Please see the section of the supplement on the astronomical applications of the pulverizer. It seems that the first problem to be solved in this case is:

$$y' = \frac{12x' - 154168}{210389},$$

where $154168 \simeq \frac{1319 \times 210389}{1800}$. Using similar rules of proportions as those described before, the value found for x", ($\frac{122727 \times 154168}{210389} = 89931 + \frac{82977}{210389}$; $x'' = 82977$) corresponds to the residue of

                                         [Degree-pulverizer]

An example for the residue of degrees (*bhāgaśeṣa*):

10. revolutions, signs and degrees are all carried away by the wind
Five minutes are seen crossed by the sun|
Say   the amount of days ⟨elapsed in the *Kaliyuga*⟩ if you know the
  ***Āśmakīya* (i.e. the *Āryabhaṭīya*)**
And also the revolutions etc, elapsed for the sun (*dinabharttr*), immediately‖

$$\begin{matrix} 0 \\ 0 \\ 0 \\ 5 \end{matrix}$$

Setting down:

The residue of degrees [obtained] is: 17532^{364}.

The result is, as before, the number of days ⟨elapsed in the *Kaliyuga*⟩ and the quotient: $\dfrac{62715}{61812}\ ^{365}$.

On the other hand, when one has performed a pulverizer with one as subtractive, a Rule of Three is performed, then also that very number of days ⟨elapsed in the *Kaliyuga*⟩ and that very quotient ⟨are obtained⟩. It is as follows: With one as subtractive, the pulverizer and the quotient should be ⟨produced⟩. And these are^{366}: $\dfrac{59873}{51011}$.

With this, the result, as before, by means of a Rule of Three is exactly ⟨the same⟩ number of days ⟨elapsed in the *Kaliyuga*⟩ and the quotient.

When the quotient is divided by three-hundred and sixty ⟨and the remainder by thirty⟩, the elapsed revolutions, the signs and the degrees result: 171, 8, [12].

On the other hand, others , having performed the pulverizer of thirty and the reduced ⟨number of⟩ terrestrial days ⟨in a *yuga*⟩^{367}, with a Rule of Three compute the elapsed degrees and the residue of signs. It is as follows:

---

revolutions. The problem then to be solved is:

$$y = \frac{576x' - 82977}{210389}.$$

^{364} $17532.4 \simeq \frac{5 \times 210389}{60}$. The value given here is an approximation.

^{365} The following pulverizer is considered, $576 \times 360 = 207360$ being the number of minutes crossed by the sun in a *yuga*: $y = \frac{207360x - 17532}{210389}$. The results obtained are the two values set down by Bhāskara. $x = 62715$ (elapsed days), $y = 61812$ (elapsed degrees).

^{366} The value given in the edition seems to be 51101, but the value given in the table p.335 is 59011.

^{367} The pulverizer considered here is:

$$y = \frac{30x - 1}{210389}.$$

[Setting down]: $\dfrac{30}{210389}$.

[In this case], the pulverizer and the quotient are: $\dfrac{7013}{1}$

When one has performed a Rule of Three with that, the residue of signs[368] and the elapsed degrees[369] are: $\dfrac{84740}{12}$.

Because the computation of the number of days ⟨in the *Kaliyuga*⟩ with that residue of signs has already been mentioned[370], it is not stated.

[Minute-pulverizer]

p.141, line 1

In the same way when one has seen the residue of minutes (*liptāśeṣa*), a pulverizer is performed. It is as follows:

An example:

11. The revolutions, signs (*ṛkṣa*), degrees (*lava*) and minutes have been taken away by the wind. [A second is seen.]‖
Let the number of days and the passed revolutions, signs (*gūha*), degrees (*aṃśa*), and minutes of the sun ⟨in the *Kaliyuga*⟩ be told‖

Setting down: $\begin{matrix} 0 \\ 0 \\ 0 \\ 0 \\ 1 \end{matrix}$.

The result, as before, is the residue of minutes[371]: 3506.

However, the results sought for a first pulverizer are those of:

$$v = \frac{30u - 17532}{210389},$$

which are found from the first with a "Rule of Three". u is the residue of signs, v the elapsed degrees.

[368] 84740 is the remainder of the division $\frac{7013 \times 17532}{210389}$.
[369] 12 is the residue of the division $\frac{1 \times 17532}{30}$.
[370] Reading *abhihitatvān* instead of the *abhihitvān* of the printed edition.
[371] 3506 is an approximation of the quotient $\frac{1 \times 210389}{60}$, 60 being the number of seconds in a minute.

Procedure: When one has multiplied the reduced revolutions of the sun ⟨in a *yuga*⟩ by zero-zero-the cube of six $(21600)^{372}$, the placement (*sthapanā*) is^{373}:

  12441600
    210389   ·

In this case, having divided the dividend by the divisor, one should place the quotient separately ⟨and keep it⟩ unerased. When one has performed the pulverizer of the terrestrial days and the residue, when one has multiplied separately the higher quantity of the ⟨two⟩ obtained by the pulverizer of the ⟨quantity⟩ kept unerased, one should add the quotient ⟨which stands below⟩. ⟨This⟩ is the quotient.^{374}

The result with this method (*krama*) is the number of days ⟨elapsed in the *Kaliyuga*⟩ and the quotient of the division. Placement: $\dfrac{125342}{7412246}$ ·

When the quotient is divided by zero-zero-the cube of six (21600) ⟨and the successive remainders by 1800 and 60⟩, the passed revolutions, signs, degrees and minutes are:

        343
          1
        27   ·
        26

Or else, when one has performed the pulverizer with one as subtractive, a Rule of Three is performed. It is as follows: With one as subtractive, the pulverizer and the quotient are: $\dfrac{81647}{4828291}$ ·

Because the remaining has been mentioned, it is not stated.

p.142, line 1 Or else, when one has performed the pulverizer of the terrestrial days and sixty and with one as subtractive, the residue of degrees and the elapsed minutes are obtained by means of a ⟨rule of⟩ proportion (*anupāta*). It is as follows: With one as subtractive, the pulverizer of the terrestrial days and sixty, and, the quotient are: $\dfrac{108701}{31}$. In the same way, the ⟨pulverizer procedure⟩ should be applied for the residues of minutes and of seconds (*tatparās*) also.

Now when someone pointing to the ⟨longitude of⟩ sun asks: "In what time will the sun be of the same ⟨longitude⟩ again?" This should be said: By ⟨the time⟩ equal to the reduced terrestrial days. Because, when the reduced terrestrial days

---

^{372}The number of minutes in a revolution.

^{373}The pulverizer procedure is performed to resolve the following problem:

$$y = \frac{(576 \times 21600)x - 3506}{210389} = \frac{12441600x - 3506}{210389}.$$

^{374}Please see the section of the supplement on the alternative procedures given for a pulverizer without remainder.

are added[375] ⟨to the given time⟩, the sun will be in the same ⟨longitude⟩[376].

[Week-day pulverizer]

Now someone pointing to ⟨the longitude of⟩ the sun on Sunday asks: "In what time will the sun on Sunday or on Mo(on)-day (*somadina*[377]) or on the week day of another planet be of the same ⟨longitude⟩ again?" It is as follows:

A pulverizer should be performed for the residue (*avaśiṣṭa*) of the division by seven of the reduced terrestrial days. When one has chosen a subtractive ⟨term for the pulverizer⟩ by means of a one-by-one increase beginning with the weekday which is immediately after the indicated week-day, what is obtained in this way is the pulverizer which is the multiplier of the reduced terrestrial days[378]; when one has added the passed number of days ⟨in the *Kaliyuga*, obtained with⟩ the indicated sun, to the reduced terrestrial days multiplied by that ⟨pulverizer⟩, the time equal to what has been produced should be announced ⟨as the answer⟩.[379]

An example: 15

> **12. In Sagittarius[380] (*dhanvin*), degrees (*aṃśa*) equal to the square of
> five (*śara*), and the square of six minutes (*maurika*), and increased
> by ten seconds (*vikalā*), ⟨these⟩ describe the mean ⟨longitude⟩ of the
> sun (*bhānu*)|**
>
> **On a Wednesday. Say clearly, in how much time the same ⟨longitude of
> the⟩ sun ⟨will be observed⟩ on Wednesday, Thursday and Friday.‖**   20

---

[375]Reading *kṣiptaiḥ* rather than *kṣiptaḥ*.

[376]This may be a reference to the same computation described p.139 line 10 (a comment in the resolution of example 9), which has been described in the Section treating of the alternative procedures of the pulverizer without remainder in the supplement.

[377]Please see the supplement for the Glossary giving a list of the name of week-days used in this commentary.

[378]Since

$$\frac{210389}{7} = 30055 + \frac{4}{7}$$

the problem to be solved by a pulverizer is:

$$y = \frac{4x - a}{7},$$

where a is the number of week-days separating the day of the week for which the sun's mean longitude is given (V), and the day of the week on which the sun at sunrise has the same mean longitude (V_a; the first week-day being excluded, the last one included).

[379]This computation is discussed in the section of the supplement on the astronomical applications of the pulverizer.

[380]Sagittarius is the ninth sign. Therefore 8 signs have been crossed by the sun when he is in Sagittarius.

On a Wednesday this is the mean ⟨longitude⟩ of the sun:
$$\begin{array}{c} 8 \\ 25 \\ 36 \\ 10 \end{array}$$

25   With this ⟨longitude of the⟩ sun, the number of days obtained with the previous procedure is 1000.[381] On ⟨the day indicated by⟩ this number of days, a Wednesday ⟨has been obtained⟩.[382]

Now a pulverizer computation. When the reduced number of terrestrial days is p.142, divided by seven, the remainder is 4. The pulverizer with one for subtractive line 1 aiming at ⟨the number of days elapsed in the *Kaliyuga* falling on⟩ Thursday is 2; [383] the reduced number of days multiplied by that produces 420778, when this is [increased by what was previously obtained, what is produced] is the time 421778.[384] In order ⟨to compute the number of days elapsed in the *Kaliyuga* falling on⟩ Friday two is subtracted in the pulverizer. What is obtained, as before, is the time 842556.[385] The pulverizer aiming at ⟨the number of days elapsed in the *Kaliyuga* falling on⟩ Wednesday is 7, and the time 1473723.[386]

---

[381]That is, having deduced from this longitude, the "residue of revolutions", 155222, a solution of:
$$y = \frac{576x - 155222}{210389},$$
is (1000,2). In other words, $A_V = 1000$.

[382]Ab.1.4d states [Sharma-Shukla 1976; p.6]:

> *budhāhnyajākordayācca laṅkāyām‖*
> (These revolutions commenced) at the begining of the sign Aries on Wednesday at sunrise at Laṅkā (when it was the commencement of the current *yuga*).

As 1001 is divisible by seven, when a 1000 days have past, the new day is once again a Wednesday.

[383]A solution of
$$y = \frac{4x - 1}{7}$$
is (2,1).

[384]With the same notation as in the supplement:
$$A_{V_a} = 210389 \times 2 + 1000 = 420778 + 1000 = 421778.$$

[385]A solution of
$$y = \frac{4x - 2}{7}$$
is (4,2).
$$A_{V_a} = 210389 \times 4 + 1000 = 842556.$$

[386]A solution of
$$y = \frac{4x - 0}{7}$$
is (7,4).
$$A_{V_a} = 210389 \times 7 + 1000 = 1473723.$$

In this way, on precisely all week-days, the time and the pulverizer should be used
cleverly (*yuktyā*).

[A different planet-pulverizer]

p.143,
line 6

And a pulverizer with an additive remainder (*upacayāgra*) must also be applied in
⟨problems involving⟩ residues of signs, degrees, and minutes.

An example:

> **13. The ⟨revolutions⟩ accomplished ⟨by the sun⟩, together with the signs
> and degrees have been taken away by the wind. The minutes to be
> crossed by the sun (*divasakara*) are seen|**
> **To measure the square of five (*viṣaya*) increased by eleven (*śiva*). Now,
> let the number of days ⟨elapsed in the *Kaliyuga*⟩ and the sun⟨'s
> longitude⟩ be stated‖**

$$\text{Setting down:} \quad \begin{matrix} 0 \\ 0 \\ 36 \end{matrix}$$

In this case when one has made into degrees the reduced revolutions, what results,
from a pulverizer together with the additive remainder[387] is, as before, the number
of days ⟨elapsed in the *Kaliyuga*⟩ and the quotient of the division, $\frac{66027}{65077}$. In
this case the quotient is produced greater ⟨than the crossed degrees⟩ by one.[388]
When one has removed one, and when the remainder is divided by three hundred
and sixty, the revolutions, signs, and degrees of the sun should be answered.

Also, when one has performed a pulverizer with one unit as additive, a Rule of
Three, with the remainder to be crossed, 126233, ⟨should be performed⟩. With
that as well what has been obtained is the number of days ⟨elapsed in the *Kaliyuga*⟩
and the quotient of the divisor, just as it has previously been written. In this way,
a pulverizer should be used for other planets ⟨also⟩.

---

[387]The pulverizer procedure is applied, here, to solve the problem:

$$y = \frac{207360x + 126233}{210389},$$

where $126233 \simeq \frac{36 \times 210389}{60}$ is the residue of degrees; and $207360 = 576 \times 360$ is the reduced
number of degrees crossed by the sun in a *yuga*.

[388]As explained in the supplement, and previously by Bhāskara, the remaining minutes given
in the problem, are those *after which* the total number of degrees crossed *will* be an integer.
Therefore, the last degree counted here, in reality, is not crossed entirely. It should be subtracted
from the number of crossed degrees found.

[A particular week-day pulverizer]

Now someone asks, in this way: "On a Sunday or on a Monday (*somadina*), the sun and the moon (*candramas*) are of that amount (*saṅkhyā*). In how much time will both be of just this amount?"

In this case, the pulverizer method (*krama*) ⟨is as follows⟩: Some quantity when divided by the reduced number of terrestrial days ⟨in a *yuga*⟩ for the sun, has a zero-remainder (*śūnyāgra*), just that ⟨same quantity when divided by the reduced number of civil days in a *yuga*⟩ for the moon too has a zero-remainder. In this example (*uddeśana*), ⟨there is⟩ an association (*sambandha*) of both of them, therefore the product of ⟨one reduced day by the quotient of the other by the quantity⟩ having such remainder for two divisors (*dvicchedāgrasaṃvargo*) has the name "procedure of equalizing (*sadṛ'sīkaraṇaṃ*) for two quantities". And, in this case, the reduced ⟨number of civil⟩ days ⟨in a *yuga*⟩ for the sun is reduced by this, 3449,[389] the result is 61. For the moon also, what results with that very reducer (*apavartana*) is 625. Hence the reduced (*apavartita*) days for the moon (2155625) are multipliers of the reduced[390] days for the sun (61). And this has just been written here. When multiplied, what results is 131493125. For the moon also when multiplied by the reduced (*apavartita*) days for the sun, what results is 131493125. [391] As before, a planet-pulverizer is to be applied with that quantity. It is as follows:

An example:

14. **Both the sun and the moon are seen by me, accurately, in the Balance-holder (*tulādharanara*) ⟨sign[392]⟩ with respectively twelve and two degrees, when ⟨the sun is⟩ rising on a Sunday|**
   And with respectively one (*śaśin*)-zero(*śūnya*)-four(*sāgara*) minutes. Moreover, in how many days will they both be of the same ⟨longitude

----

[389]The greatest common divisor of the number of revolutions of the moon in a *yuga* (57753336) and the number of civil days in a *yuga* (1577917500) is 732. Thus for the moon, the reduced number of lunar revolutions is $\frac{57753336}{732} = 78898$ and the reduced number of terrestrial revolutions is $\frac{1577917500}{732} = 2155625$. 3449 is the greatest common divisor of the reduced number of civil days in a *yuga* for the sun (210389) and the reduced number of civil days in a *yuga*, for the moon (2155625). And

$$210389 = 61 \times 3449,$$
$$2155625 = 625 \times 3449.$$

[390]Reading *nirapavartita* rather than the *niravartita* of the printed edition.
[391]
$$131493125 = 210389 \times 625,$$
$$131493125 = 2155625 \times 61.$$

This number is the LCM of 210389 and 2155625.
[392]Balance is the 7th sign, therefore six signs have been crossed, by the two planets.

again⟩ **successively on a Thursday, then on a Friday, and on a Saturday?‖**

One should know that the ⟨longitude of the⟩ sun is greater by seven (*bhūdhara*)-one (*indu*) seconds.|

One should subtract seconds equal to eighteen (*dhṛti*) from the ⟨longitude of the⟩ moon (*niśānātha*)‖

	[sun]	[moon]
	6	6
Setting down[393]:	12	2
	1	39
	17	42

The result, from both, is the number of days ⟨elapsed in the *Kaliyuga*⟩: 7500. [394]

Procedure: The remainder of the division of the reduced days for the sun and the moon (131493125) by planet ⟨days⟩ (i.e. week days) (18784732[395]) is $\frac{1}{7}$. The fourth ⟨day⟩ after the indicated week-day is Jupiter (Thursday), the fifth ⟨day⟩ is Venus (Friday), the sixth ⟨day⟩ is Saturn (Saturday). With these, in this way these are obtained from the residue of the division by planet ⟨days⟩ (i.e. week days) in due order: for Jupiter (Thursday) 4, for Venus (Friday) 5, for Saturn (Saturday)

---

[393]The 39 minutes and 42 seconds of the moon, correspond to the 40 minutes decreased by 18 seconds.

[394]By solving by a pulverizer procedure either one of the problems stated below, the value found for x corresponds to the number of days elapsed in the *Kaliyuga* on that Sunday:

$$y = \frac{576x - 112219}{210389},$$

where $\frac{691277 \times 210389}{1296000} \simeq 112219$ is the residue of revolutions. $((30 \times 60 \times 60 \times 6) + (60 \times 60 \times 12) + (60 \times 1) + 17 = 691277$ is the longitude of the sun reduced to seconds.

Or

$$y = \frac{78898x - 1093750}{2155625},$$

where $\frac{657582 \times 2155625}{1296000} \simeq 1093750$ is the residue of revolutions. $((30 \times 60 \times 60 \times 6) + (60 \times 60 \times 2) + (60 \times 39) + 42 = 657582$ is the longitude of the moon reduced to seconds.

[395]I do not know where this value is derived from, thought in some way from the number of weeks in a *yuga*. Obviously, $131493125 = 18784732 \times 7 + 1$.

6.[396] Just these are the multipliers of the reduced days for the sun and the moon. When one has added the number of days obtained into ⟨that⟩ multiplied by the multipliers, in due order ⟨the result is⟩[397] : [525980000, 657473125, 788966250 days].

[A pulverizer using the sum of ⟨the longitudes of⟩ two planets]

Now, when someone, having put together (i.e. summed, *ekatra kṛtvā*) ⟨the mean longitudes of⟩ two planets asks the number of days ⟨elapsed in the *Kaliyuga*⟩, this is a method (*upāya*) for that ⟨question⟩: When one has performed a reduction of, the ⟨number of⟩ terrestrial days ⟨in a *yuga*⟩, and of the sum of the ⟨number(s) of⟩ revolutions ⟨in a *yuga*⟩ of the indicated planets, a pulverizer should be performed. It is as follows:

p.145, An example :
line 1

> **15. The ⟨longitudes of⟩ the sun and moon, added together, have been**
> **seen to be thirty ⟨minutes⟩ and five ⟨degrees⟩ and one (*śaśāṅka*)**
> **⟨sign⟩ |**
> **Tell the amount of days (*dinarāśi*) elapsed ⟨in the *Kaliyuga*⟩ and the**
> **passed revolutions (*cakra*)‖**

5   Setting down: $\begin{matrix} 1 \\ 5 \\ 30 \end{matrix}$

Procedure: ⟨The sum of⟩ the revolutions of the sun and the moon ⟨in a *yuga*⟩[398] is 62073336. [And the number of terrestrial days in a *yuga* is] 1577917500. Both reduced by twelve produce: $\dfrac{5172778}{131493125}$

---

[396] Please see supplement on the names of planets, which has a section on the naming of Weekdays. The problem to be solved by a pulverizer, for Thursday, is:

$$w = \frac{1 \times v - 4}{7}$$

A solution has been obtained for $v = 4$. The problem to solve for Friday is:

$$w = \frac{1 \times v - 5}{7}.$$

A solution is obtained for $v = 5$. The problem to solve for Saturday is:

$$w = \frac{1 \times v - 6}{7}.$$

A solution is obtained for $v = 6$.

[397] $\begin{aligned} 7500 + 131493125 \times 4 &= 525980000 \\ 7500 + 131493125 \times 5 &= 657473125 \\ 7500 + 131493125 \times 6 &= 788966250 \end{aligned}$

[398] According to the values given in Ab.1.3.

The residue of revolutions[399]: 12966683. In this case, the result as before is the 10
number of days ⟨elapsed in the *Kaliyuga*⟩ and the quotient[400]: $\dfrac{57942886}{3459564}$.

When one has performed ⟨a pulverizer⟩ with one as subtractive, too, with a Rule
of Three, that very number of days ⟨elapsed in the *Kaliyuga*⟩ and the quotient are
obtained[401]. It is as follows: The pulverizer with one unit as subtractive and the
quotient are : $\dfrac{57699692}{2269835}$. 15

[In this case also, as it has been said previously, with a Rule of Three ⟨that⟩ very
number of days ⟨elapsed in the *Kaliyuga*⟩ and the quotient ⟨are obtained⟩.]

In this way, in questions concerning the sums of other ⟨planets⟩ as well, a pulverizer
is to be performed, and also ⟨in questions⟩ concerning residues of signs, degrees
and minutes. In this very way, in the case of the sums of three or four ⟨planets⟩
too an explanation should be given in detail ⟨if necessary⟩.

[A pulverizer with two remainders] 20

Now, when pointing at ⟨the longitude of⟩ a planet (*divicara*) produced at the time
when the sun completes what remains of a revolution, someone asks the ⟨number
of⟩ revolutions ⟨performed⟩ by ⟨that planet⟩, this is a method for that ⟨question⟩:
When one has reduced the ⟨number of⟩ revolutions ⟨performed⟩ by a planet ⟨in
a *yuga*⟩ and the ⟨number of⟩ revolutions ⟨performed⟩ by the sun ⟨in a *yuga*⟩, a
pulverizer should be applied. It is as follows:

An example:

> **16. The desired ⟨longitude of⟩ Mars (*medinīhṛdayaja*), produced at the**
> **time when the sun (*bhānu*) completes a revolution is |** 25
> **Two, three times five and five (*viṣaya*) in houses (i.e. signs), etc. Say**
> **the passed revolutions of Mars (*kuja*) and the sun (*arka*).‖**

Setting down : $\begin{array}{c} 2 \\ 15 \\ 5 \end{array}$ p.146,
line 1

---

[399] $12966683 \simeq \frac{2130 \times 131493125}{21600}$, where 2130 is the mean longitude of the sun and moon given in
the above problem, reduced to minutes.

[400] The problem to be solved by a pulverizer here is

$$y = \frac{5172778x - 12966685}{131493125},$$

where x is the number of terrestrial days elapsed since the beginning of the *Kaliyuga* at that
time, and y is the sum of the revolutions performed, since the beginning of the *Kaliyuga*, by
both the sun and the moon.

[401] Reading *labhyante* rather than *labhyate*.

5   Procedure: The revolutions of Mars and the sun ⟨in a *yuga*⟩ are[402]: $\dfrac{2296824}{4320000}$

Both are reduced by twenty four: $\dfrac{95701}{180000}$

The residue of revolutions[403] is this: 37542. The results are the passed revolutions of the sun (*ravi*) and Mars[404], 68142, 36229.

10   When one has performed a pulverizer with one as subtractive, also, just those revolutions ⟨are produced⟩ with a Rule of Three. It is as follows: The pulverizer with one as subtractive and the quotient are: $\dfrac{174301}{92671}$

It is exactly in this way also for other planets.

Or else when ⟨someone⟩ pointing at a planet asks ⟨the number of passed revolutions⟩ of precisely another planet, then again a pulverizer should be performed by

15   choosing ⟨an appropriate⟩ divisor and dividend.

Now, having indicated ⟨the mean longitude of⟩ a planet in revolutions, etc. produced at a time different from the time when ⟨the sun⟩ completes a revolution, ⟨someone⟩ asks, then, from the ⟨elapsed⟩ revolutions as they are[405], the revolutions, signs, degrees and minutes of the sun; this is a method (*upāya*) for that pulverizer computation (*kuṭṭākārānayana*) too: When one has reduced the revolutions of the sun ⟨in a *yuga*⟩ multiplied by zero (*kha*)-zero-the cube of six (*ṣaḍgaṇa*) (21600) together with the indicated ⟨planet's number of⟩ revolutions ⟨in a *yuga*⟩ a pulverizer operation (*kuṭṭākāravidhi*)⟨should be performed⟩. It is as follows:

20   An example :

> **17. Jupiter (*adhirūḍhamahendrasūrau*[406]) is in half the degree of his own apogee (*ucca*[407]) and, hence, ⟨someone⟩ in the line of the *Āśmaka* (i.e. Āryabhaṭa) asks the ⟨degrees, etc⟩ crossed by the sun|**
> **Who has purified by the expansion of his energy the faces of the cardinal directions, what is the answer? Say ⟨it⟩ quickly to that ⟨person⟩, O one of great intellect!‖**

---

[402]According to Ab.1.3.

[403]$37542 \simeq \frac{4505 \times 180000}{21600}$, where 4505 is the longitude of Mars given in the problem, reduced to minutes.

[404]The pulverizer procedure is thus applied to the following problem:

$$z = \frac{95701y - 37542}{180000},$$

where y is the number of revolutions accomplished at that date by the sun and z the integral number of revolutions accomplished by Mars.

[405]Reading as in all manuscripts *tad yathā bhagaṇāt* rather than *te tathā* of the printed edition.

[406]This compound is somewhat difficult to interpret, please see the Appendix of the glossary giving the names of planets.

[407]For a definition of the *ucca*, please see the Appendix d. The *ucca* of Jupiter, according to K.S. Shukla quoting the *Bṛhajjātaka* [Shukla 1976; p. 322], is in the fifth degree of Cancer (which is the fourth sign).

Setting down : $\begin{array}{c} 0 \\ 3 \\ 4 \\ 30 \end{array}$ 25

Procedure: When [the revolutions of Jupiter] ⟨in a *yuga*⟩ , together with the revo-   p.147,
lutions of the sun ⟨in a *yuga*⟩ multiplied by zero-zero-the cube of six (21600), are   line 1
reduced by a hundred and ninety two, what results is[408]: $\dfrac{1897}{486000000}$.

The residue ⟨of revolutions⟩[409] which is a subtractive is: 127575000.

The result is the sun's crossed revolutions[410]: 78975000. The minutes divided by   5
zero-zero-the cube of six are the revolutions, signs, degrees and minutes of the sun
$\begin{array}{c} 3656 \\ 3 \\ 0 \quad \cdot \\ 0 \end{array}$

Having performed a pulverizer with one as subtractive also, with a Rule of Three,   10
just that ⟨amount of elapsed revolutions⟩ is obtained. It is as follows: The pulverizer
with just one for subtractive and the quotient are: $\dfrac{135014233}{527}$.

In this way ⟨a pulverizer⟩ should be applied also ⟨for problems⟩ in residues of signs
etc.

[Time-pulverizer]

Now, the time-pulverizer (*velākuṭṭākāra*) ⟨is explained⟩. When someone pointing   15
at ⟨the mean longitude of⟩ a planet produced at a time different from sunrise,
asks the number of days ⟨elapsed in the *Kaliyuga*⟩ (*divasagaṇa*), this is a method
of computation (*ānayanopāya*), for that ⟨question⟩: When one has multiplied the
reduced ⟨number of⟩ days ⟨in a *yuga*, for that planet⟩ by the denominator (*chedha*)

---

[408]According to Ab.1.3. the number of revolutions of Jupiter in a *yuga* is 364224. Therefore:
$$\frac{364224}{192} = 1897$$
$$\frac{4320000 \times 21600}{192} = 486000000$$

[409]$127575000 = \frac{5670 \times 486000000}{21600}$, where 5670 is the longitude, reduced to minutes, of Jupiter as
given in the problem.

[410]The problem to be solved by a pulverizer here is:
$$z = \frac{1897Y - 127575000}{486000000},$$
where z is the integral number of revolutions crossed by Jupiter since the beginning of the
Kaliyuga and Y the number of minutes crossed by the sun during that time.

of the desired time, and brought about a pulverizer, as before, what is divided by the denominator of the desired time is the number of days ⟨elapsed in the *Kaliyuga*⟩ (*ahargaṇa*). It is as follows:

An example:

18. The desired ⟨mean longitude⟩ of the sun (*bhartur divasasya*) produced at mid-night (*rātrerdhakāla*) with half the Capricorn ⟨sign⟩ (*mṛga*), the remaining minutes which are eight times four,|
Increased by two-thirds of a minute. Quickly say, according to the ⟨teachings of⟩ the *Āśmaka* (i.e. Āryabhaṭa), the number of ⟨elapsed⟩ days[411] and the ⟨number of⟩ revolutions ⟨accomplished by the sun⟩.‖

25   Setting down[412]:
$$\begin{array}{c} 9 \\ 15 \\ 32 \\ 40 \end{array}$$

Procedure: Since the number of ⟨elapsed⟩ days is smaller by a quarter, four is the multiplier of the reduced days, therefore when one has reduced the revolutions of the sun ⟨in a *yuga*⟩ by a quarter [413], the placement is: $\dfrac{144}{210389}$. The residue of

p.148,   revolutions is 166876.[414] The result as before is the pulverizer [7003]. The number
line 1   of ⟨elapsed⟩ days is ⟨obtained from⟩ one fourth of that ⟨decreased by three⟩, [1750].

In the same way, an example related to sunsets (*āstamayika*).

19. All the assemblage of the series of digits (*aṅka*) beginning with the elapsed revolutions, computed according to a procedure (*krama*), of the sun, and which is beautiful when hidden by the elevated summit of the western mountain, has been forgotten.|
5
The remaining amount of minutes, being a clear quantity ⟨expressed⟩ in digits, is seen to be a hundred and three; three (*guṇa*)-zero (*viyad*)-one (*uḍupa*). Quickly, let the number of days computed in

---

[411]reading *dināni* rather than *dinādi*.

[412]Capricorn is the tenth sign, so half that sign indicates that 9 signs and 15 degrees have been crossed.

[413]Days are defined from one sunrise to another, therefore mid-night corresponds to three quarters of a day, or a whole day minus one fourth. The problem to be solved by a pulverizer here is:
$$y = \frac{576(x - 1/4) - 166876}{210389} = \frac{144X - 166876}{210389}.$$
The expression giving 4 as a multiplier of "the reduced days" is enigmatic to me. This may be a reference to the fact that $X = 4x - 1$ is considered, but it is difficult to understand why X would bear such a name.

[414]$166876 \simeq \frac{1027960 \times 210389}{1296000}$, where 1027960 is the sun's mean longitude reduced to seconds, and 1296000 is the number of seconds in a revolution.

the *kaliyuga* **and the** ⟨**number of elapsed**⟩ **revolutions etc. of the sun
be told.**‖

Setting down[415]: $\dfrac{288\ [\times 21600]}{210389}$ the remainder is 103

The placement of the number of days ⟨elapsed in the *Kaliyuga*, obtained⟩, as before   10
with this subtractive is[416]: 99275.

The quotient ⟨divided by 21600, produces⟩ the crossed revolutions, signs, degrees

and minutes of the sun: $\begin{array}{c} 271 \\ 9 \\ 16 \\ 13 \end{array}$.

The pulverizer with one as subtractive and the quotient are here: $\dfrac{163294}{4828291}$. With   15
that, by means of a Rule of Three again, the previously computed number of days
and quotient of division ⟨beginning with⟩ the revolutions, etc. ⟨are obtained⟩

A mid-day example :   17

> **20. [The residue of revolutions] of the sun who has reached the zenith**
> **(*madhya*) and who illuminates the faces of the directions with the**
> **abundance of ⟨its⟩ extremely harsh ⟨rays⟩, is seen to be equal to**
> **[zero]-nine-seven (*naga*)-four-five (*bhūta*)-one (*śītāṃśu*)**‖
> **The number of days ⟨elapsed⟩ and the [amount (*caya*) of] elapsed**   20
> **revolutions established [at that very time] should be told as obtained**
> **by one who has properly learned the rules of the *Āśmaka*'s teaching**
> **on the pulverizer**‖

Setting down[417] :[$\dfrac{144}{210389}$] The residue of revolutions is 154790.

---

[415]$288 = \frac{576}{2}$: since sunset here indicates the middle of a day, defined from one sunrise to
another. The remainder of minutes is considered. It is multiplied by the number of minutes in a
revolution, 21600.

[416]The problem to be solved by a pulverizer is:

$$y = \frac{288 \times 21600X - 103}{210389},$$

where $X = 2x + 1$, using the tabulated solutions of

$$Y = \frac{288 \times 21600X - 1}{210389}.$$

For the details of the computation, see [Shukla 1976; p.323-324]. y is the number of minutes
crossed in x days, here $y = 5870773$. The value found for X is 57182401, so that $x = 99275$.

[417]The problem to be solved by a pulverizer here, since midday corresponds to a quarter of a
day is:

$$y = \frac{144X - 154790}{210389},$$

By proceeding as before with this residue of revolutions, the result is [the pulverizer] and the quotient $\frac{3997}{2}$; the number of ⟨elapsed⟩ days is one fourth of the pulverizer[418], 999.

p.149, line 1   The pulverizer with one as subtractive and the quotient are $\frac{168019}{115}$. As before, the computation of the number of ⟨elapsed⟩ days ⟨has been obtained⟩ with that by means of a Rule of Three.

Likewise, concerning the time ⟨units called⟩ *yāma*[419], *muhūrta, nāḍī, vināḍikā* also, according to circumstances, cleverly (*yuktyā*), a pulverizer should be considered. It is as follows:

5   An example :

> **21. The revolutions, etc. for the sun (*tigmāṃśu*), which come from the accumulated number of days ⟨elapsed in the *Kaliyuga*⟩ together with a certain amount of *nāḍīs*, have just been dissolved by a storm.|**
> **The residue of minutes, seventy increased by one, is seen by me. The number of days (*dyugaṇa*) ⟨elapsed⟩, the sun's elapsed ⟨revolutions⟩ and the exact ⟨amount of⟩ *nāḍikas* should be stated.‖**

10   The residue of minutes is 71.

Procedure: One should reduce together the reduced ⟨number of⟩ revolutions of the sun ⟨in a *yuga*⟩ and sixty. With one twelfth of sixty, five ⟨is obtained⟩. One twelfth of the reduced ⟨number of⟩ revolutions of the sun ⟨in a *yuga*⟩ is forty-eight. When one has multiplied by five, the ⟨number of⟩ terrestrial days ⟨in a *yuga*⟩ [and the residue], the placement is[420]:

15    $$\frac{48[\times 21600]}{1051945} \quad , \text{ the residue is } 71[\times 5]$$

---

where $X = 4x + 1$. (y,X)=(2, 3997)

[418] $x = \frac{X-1}{4} = 999$ and $\frac{X}{4} = \frac{3996}{4} = 999, 25 \simeq 999$.

[419] Please see the part of the Glossary on measure units.

[420] In this case the time elapsed since the beginning of the *Kaliyuga* has an additional fractional part (n) in *nāḍīs*. Since there are sixty *nāḍīs* in a day, $x + \frac{n}{60}$ is the time in days elapsed since the beginning of the *Kaliyuga*; and $21600y + \frac{71}{210389}$ the number of minutes crossed by the sun (this value of $\frac{71}{210389}$ thought a bit surprising, and nowhere mentioned in the text is the only one that makes sense in the following computations), then we have the following ratio:

$$\frac{x + \frac{n}{60}}{210389} = \frac{21600y + \frac{71}{210389}}{21600 \times 576},$$

or the following equation:

$$Y = \frac{(21600 \times \frac{576}{60})X - 71}{210389},$$

where $X = 60x + n$ and $Y = 21600y$. As $\frac{576}{60} = \frac{48}{5}$, this equation can be further simplified:

$$Y = \frac{21600 \times 48X - 71 \times 5}{210389 \times 5},$$

[When one has performed a reduction by five in these ⟨places⟩ the placement is
$\frac{207360}{210389}$, the residue is 71 $]^{421}$

When one has, as before, performed a pulverizer, the result is the number of days
⟨elapsed in the *Kaliyuga*⟩: 720, the *nāḍīs* are 3.

And the elapsed revolutions, etc. of the sun are: $\begin{matrix} 1 \\ 11 \\ 19 \\ 41 \end{matrix}$ 20

The pulverizer with one as subtractive also, and the quotient of the division are:
59873
59011

As before with a Rule of Three the number of days ⟨elapsed in the *Kaliyuga*⟩ and
the quotient ⟨are obtained⟩. 25

[A pulverizer with a non-reduced residue]

Now, furthermore, having indicated a residue, which precisely is non-reduced,
⟨someone⟩ asks for the ⟨elapsed⟩ number of days and the passed ⟨revolutions⟩;
this is a method of computation for that ⟨question⟩ also. When one has performed
the reduction, by a unique reducing divisor (*cheda*), of the divisor, dividend and
remainder, as before, a pulverizer is performed. Now, on the other hand, but if p.150,
that example is such that these divisor, dividend and remainder do not allow such line 1
a reduction with a unique divisor, as there is no such one quantity ⟨that satisfies
this equation⟩, ⟨such a quantity⟩ is not computed ⟨with a pulverizer⟩.

An example:

**22. The number which is the residue of revolutions of the sun (*dinakara*)
and which has been produced from the unchanged terrestrial days
and revolutions ⟨of the sun in a *yuga* ⟩ |** 5
**Is five (*śara*)-two (*yama*)-eight (*vasu*), with a hundred for multiplier.
Say the amount of days and the revolutions obtained with that
⟨quantity.⟩ ||**

Setting down: $\frac{4320000}{1577917500}$, the residue: 82500

$210389 \times 5 = 1051945$.

[421] This placement is added by the editor. It points out that 21600 is also divisible by five, so
an additional simplification is possible, though we do not know if it was actually carried out.

And when one has reduced these quantities[422] by zero (*kha*)-zero (*ākāśa*)-five (*śara*)-seven (*muni*), with a pulverizer, the number of days and the elapsed revolutions of the sun ⟨in the *Kaliyuga* are obtained⟩ [423]: $\dfrac{199066}{545}$.

[A particular pulverizer with two remainders]

Now someone asks a single number of days ⟨elapsed in the *Kaliyuga*⟩, with the method (*nyāya*) of "⟨a quantity that has such⟩ remainders for two divisors" (*[dvi]-chedāgra*), for two different residues of revolutions for two planets; the computation of that "⟨quantity that has such⟩ remainders for two divisors" for that ⟨question is performed⟩ with this ⟨rule:⟩ "One should divide the divisor of the greater remainder".

An example:

23. A certain amount of days is divided ⟨separately⟩ by the ⟨reduced number of⟩ days ⟨in a *yuga*⟩ for Mars (*aṅgāraka*) and for the Sun,

[In this case, I do not know] the quotients, nor have I observed their remainders|

These two ⟨remainders⟩ are multiplied by ⟨their respective reduced numbers⟩ of revolutions ⟨in a *yuga*⟩ and then divided respectively by their ⟨reduced numbers of⟩ days ⟨in a *yuga*⟩

In this case, the result is blown away [by the wind and at that time their remainders re]main. ‖

The remainder for the Sun is two (*aśvin*)-seven (*naga*)-four (*abdhi*)-eight (*nāga*)-three (*śikhin*) and for Mars (*kuja*) it is said to be

five (*bhūta*)-two (*aśvin*)-six (*aṅga*): zero (*nabhas*, the sky)-eight-one (*śītakiraṇa*)-seven (*kṣoṇīdhara*)-seven (*kṣmābhṛt*)|

Having computed for these two, separately, the elapsed number [of days ⟨in the *Kaliyuga*⟩ for the Sun and Mars] and the remainder for both, O mathematicians, you should narrate ⟨them⟩ in due order‖[424]

Setting down: For the sun   38472
               For Mars (*bhauma*)   77180625

p.151, line 1   When one has performed a pulverizer with these two residue of revolutions, by means of the previous procedure, separately; the two number of days ⟨elapsed in

---

[422]Reading the plural accusative *etā rāśīn* rather than the plural nominative *ete rāśayaś* of the printed edition.

[423]After reduction by 7500, which is the Greatest Common Divisor of (4320000, 1577917500), the problem to be solved by a pulverizer considered here is:

$$y = \frac{576x - 11}{210389}.$$

the *Kaliyuga*⟩ obtained are, for the Sun, 8833; for Mars, 640000.[425] A pulverizer, with these two remainders, is performed by means of that ⟨rule:⟩ "One should divide the divisor of the greater remainder by the divisor of the smaller remainder".

And thus the setting down is[426]:   [for the Sun]   8833   [for Mars]   640000
                                                         210389   131493125

The difference of remainders is 631167.

When one has performed a reduction of the difference of the remainders, and of both divisors, by this 210389, the placement is[427]: $\dfrac{1 \quad 3}{625}$

In this case, the divisor which is a small number is one, therefore the entire quantity should be subtracted ⟨from the pulverizer⟩, hence ⟨the procedure⟩ is explained by inverting the quantities. And then the setting down is 1. Here there is a unit. "This quantity which is one, multiplied, by what is ⟨sought⟩, decreased by the

---

[424]The problem stated here, is in other words: After reduction, the number of civil days in a *yuga* for the Sun is 210389, and the number of its revolutions is 576. For Mars they are respectively 131493125 and 191402. Let N be a whole number (called "a certain amount of days") such that

$$\begin{cases} \dfrac{N}{210389} = q_s + \dfrac{r_s}{210389} \\[2mm] \dfrac{N}{131493125} = q_m + \dfrac{r_m}{131493125} \end{cases}$$

If r_s and r_m are known, such a problem can be solved by a pulverizer procedure, as described in Bhāskara's first interpretation of verses 32-33 ([Shukla 1976; p.132-134]). In this problem, however, they are not known, but, the following information is given:

$$\begin{cases} \dfrac{576r_s}{210389} = q'_s + \dfrac{38472}{210389} \\[2mm] \dfrac{191402r_m}{131493125} = q'_m + \dfrac{77180625}{131493125} \end{cases} \Leftrightarrow \begin{cases} \dfrac{576r_S - 38472}{210389} = q'_S \\[2mm] \dfrac{191402r_M - 77180625}{131493125} = q'_M \end{cases}$$

38472 and 77180625 can be seen as residues of revolutions, r_s and r_m as the number of days elapsed in the *Kaliyuga* for each planet, and q'_s and q'_m as the number of revolutions performed during that time.

[425]In other words, solving the last set of equations, what is found is: $r_s = 8833$ and $r_m = 640000$.

[426]In other words N can be found considering:

$$\begin{cases} N = 210389q_s + \dfrac{8833}{210389} \\[2mm] N = 131493125q_m + \dfrac{640000}{131493125} \end{cases}$$

or

$$q_M = \frac{210389q_S - (640000 - 8833)}{131493125}.$$

Bhāskara, ambiguously quotes the beginning of the verse which may be interpreted in any of the two ways. In fact he sets down a pulverizer with remainder and solves the problem according to this procedure. However, in an intermediary step, he sets down a "pulverizer without remainder".

[427]After reduction the problem considered is:

$$q_m = \frac{1q_s - 3}{625}$$

difference of remainders which is three and [when divided by five -two-six (625)]
gives an exact division." The quantity obtained is three units. When, in the order
of the inverse quantities, this is divided by five (*śara*)-two (*yama*)-six (*r̥tu*) (625)
the remainder which is the pulverizer and which is three is multiplied by that
⟨reduced⟩ divisor which is a large number; what results is 1875.[428] That divisor
of the greater remainder, 131493125 is multiplied by that and increased by the
greater remainder; what is produced is ⟨the quantity which has such⟩ remainders
for two divisors, five (*śara*)-seven (*adri*)-three (*guṇa*)-nine-four (*abdhi*)-two-zero
(*viyat*)-five (*bhūta*)- five (*śara*)-six (*rasa*)-four (*abdhi*)- two (*netra*); setting down
with digits also 246550249375.[429] This is the quantity which has such remainders
for two divisors. In this way, when performing a pulverizer with another remainder
for ⟨another⟩ divisor, the product of the two divisors multiplied by the remainder
for two divisors becomes a divisor (*hāratā*). The computation of the remainder
for three divisors ⟨has been obtained⟩ by means of a pulverizer with ⟨the third
quantity⟩ quantity and the remainder for two divisors.[430] In the same way, the
⟨computations for⟩ the remainder for four ⟨divisors⟩ should be understood by
means of own's one intellect.

p.151,   [A pulverizer with two remainders when orbital operations ⟨occur⟩]
line 21

Now an example when the number of days ⟨elapsed in the *Kaliyuga* is computed⟩
with orbits (*kakṣyā*):

> **24. The remainders ⟨of the revolutions of⟩ the sun and the moon com-
> puted with the numbers ⟨obtained⟩ from an orbital method for the
> Sun and the moon are, for the Sun**
> **Two-eight-five (*iṣu*)-four (*abdhi*[431])-four (*kr̥ta*)-four (*abdhi*)-zero (*kha*)-
> five (*iṣu*)-three (*bhuvana*)-nine (*chidra*)-one (*indu*) is mentioned|**
> **⟨For the Moon⟩ it is equal to nine (*nanda*)-six (*aṅga*)-two (*aśvin*)-one
> (*niśākara*) multiplied by the square of a thousand.**
> **That which has ⟨such⟩ remainders for two ⟨divisors⟩, the number of
> days and their revolutions crossed through the *Kalibhuj* should be
> stated‖**

---

[428]I am not quite sure what is intended by Bhāskara here. A solution of the above last equation
found by trial and error would be $(q_s, q_m)=(3, 0)$. It seems that because if $q_S = 3$, the numerator
becomes zero, the value accepted for q_m is $3 \times 625 = 1875$ on account of an inverse procedure.
As it is specified in a footnote at the beginning of the section of the supplement explaining the
procedure for the pulverizer with remainder, if one divisor is equal to 1, then the quotient of the
second divisor considered can be any given integer. So that any value adopted here for q_m gives
a final integral result.

[429]Considering 1875 as a value for q_M, we have $N = 131493125 \times 1875 + 640000$.

[430]In other words if there were a third couple with a divisor C and a remainder R, one should
apply the pulverizer computation to the two couples formed by (C, R) and $(210389 \times 131493125,$
246550249375).

[431]Adopting this reading rather than the printed: *adhdhi*.

$$19350444582 \qquad \text{for the Sun}$$
$$49797813966$$

Setting down [432]:

$$1269000000 \qquad \text{for the Moon}$$
$$3724920000$$

The difference of the remainders is 18081444582 . These give a reduction of the [5] quantities for ⟨the difference⟩ of remainders and the divisor. When one has reduced by four (*veda*)-nine-six (*ṛtu*)-zero-two (*yama*) [20694] the placement ⟨is as follows⟩:

---

[432] From rules given in the *Āryabhaṭīya*, that we have exposed in the Appendix giving some elements of hindu astronomy, we know that the orbit of the sun is:

$$\frac{12474720576000}{4320000} = 288766 + \frac{3456000}{4320000} = 2887666, 8 \ yojanas,$$

and that the orbit of the moon is

$$\frac{12474720576000}{57753336} = 216000 \ yojanas.$$

From these, the following relation between the mean longitudes of the sun (λ_S) and the mean longitude of the moon (λ_C), and the number of days elapsed since the beginning of the *Kaliyuga* is known, from a brief statement of Bhāskara at the end of this commentary, and also a verse of the *Mahābhāskarīya*, given in the above mentioned supplement:

$$\lambda_S = \frac{12474720576000x}{1577917500 \times 2887666, 8}.$$

$$\lambda_C = \frac{12474720576000x}{1577917500 \times 216000}.$$

Note that there would be an obvious simplification here, that does not seem to be carried out:

$$\lambda_S = \frac{12474720576000x}{1577917500} \times \frac{4320000}{12474720576000} = \frac{4320000x}{1577917500} = \frac{576x}{210389}$$

The previous quotients, when reduced by 91500 produce:

$$\lambda_S = \frac{136335744x}{49797813966}$$

$$\lambda_C = \frac{136335744x}{3724920000}$$

We can recognize here the divisors set down, respectively for the Sun and the moon. The residues considered in this problem would be the non-integral part of the number of revolutions performed since the beginning of the *Kaliyuga* by the sun and the moon. That is, for the Sun, the part of its revolutions that is not divisible by 49797813966; and for the Moon the part of its revolutions that is not divisible by 3724920000.

This is equivalent to:

$$49797813966\lambda_S = 3724920000\lambda_C = 136335744x = N.$$

The problem to be solved here is, at first, to find the whole number N such that:

$$\begin{cases} N = 4979781396z + 19350444582 \\ N = 3724920000y + 1269000000 \end{cases} \Leftrightarrow z = \frac{3724920000y - (19350444582 - 1269000000)}{4979781396},$$

where $z = M_S$ is the number of integral revolutions performed by the sun, and $y = M_C$ those performed by the moon.

The orbit of the sun (*ravikakṣyā*), i.e. 49797813966) reduced by this, 20694, is 2406389. ⟨The orbit⟩ of the moon, also ⟨is reduced⟩, 180000. The reduced difference of remainders is 873753[433].

The quantity obtained by a pulverizer method (*kuṭṭākāranyāya*) with these values for the divisors and the difference of remainders, is three (*guṇa*)-two (*yama*)-seven (*adri*)-three (*puṣkara*)-six (*ṛtu*)-five (*śara*)- six (*aṅga*)-seven (*adri*)-one (*indu*)-two (*yama*), and with digits also 2176563723[434]. "When divided by the divisor of the smaller remainder, the remainder is" (Ab.2.33). This quantity is divided by that divisor with a smaller remainder, 180000, the remainder is [3723[435]]. The divisor of the greater remainder is multiplied by [that] remainder, what results is seven-four (*udadhi*)-two (*yama*)-six (*aṅga*)-eight-nine (*nanda*)-eight-five (*śara*)-nine (*nanda*)- eight (*vasu*), and with digits also 8958986247. This quantity is [multiplied] by that reducer [20694, what is produced] is eight (*vasu*)-one (*indu*)-four (*udadhi*)- five (*bhūta*)-nine (*randhra*)-three (*agni*)-one (*indu*)-six (*rasa*)-two (*yama*)-seven (*adri*)-nine (*nanda*)-three (*agni*)-five (*śara*)-eighteen (*dhṛti*), and with digits also 185397261395418. Just this is increased by the greater remainder [zero-zero-zero- zero]-four (*abdhi*)-eight (*vasu*)-eleven (*rudra*)-six (*rasa*)-six (*aṅga*)-one (*indu*)-four (*udadhi*)-five (*śara*)-eighteen (*dṛti*), and with digits also 185416611840000. This is the quantity which has ⟨such⟩ remainders for two divisors[436].

[When one has made a division] of this ⟨number⟩ [by the orbit of the sky] ⟨previously⟩ reduced together with the terrestrial days ⟨in a *yuga*⟩, the number of days ⟨elapsed in the *Kaliyuga*⟩ has been obtained.[437]

Why however ⟨⟨(does one) perform⟩ a reduction of the terrestrial days and the orbit of the sky?

This is stated: In a computation of ⟨the mean longitude of⟩ planets by means of the orbits, the number of days ⟨elapsed in the *Kaliyuga*⟩ is a multiplier of the orbit of the sky, the divisor is the product of the terrestrial days ⟨in a *yuga*⟩ with its

---

[433] After reduction by 20694, the problem to be solved is:

$$z = \frac{180000y - 873753}{2406389}$$

[434] In the pulverizer procedure, in what we have called in the supplement "Step 4", which works backup the column obtained after an interrupted "Euclidean Algorithm", the quantity noted q_2'' obtained here, has such a value.

[435] Rather than the 3727 of the printed edition.

[436] In other words:

$$N = 373 \times (2406389 \times 20694) + 19350444582.$$

[437] Since

$$N = 136335744x \Leftrightarrow x = \frac{N}{136335744}.$$

As the value for N has just been computed, x is known as well.

(the planet's) own orbit[438].

⟨Therefore, the quotient⟩ obtained from ⟨the division of⟩ the orbit of the sky by the reducer of the terrestrial days ⟨in a *yuga*⟩ and the orbit of the sky, that is ⟨a division by⟩ zero (*viyad*)-zero (*ambara*)-fifteen (*tithi*)-nine (*nanda*) (91500), is four (*kṛta*)-four (*udadhi*)-seven (*naga*)-five (*śara*)-three (*rāma*)-three (*agni*)-six (*rasa*)-three (*guṇa*)-one (*indu*), and with digits also 136335744. And from the terrestrial days ⟨in a *yuga*⟩, five (*śara*)-four (*abdhi*)-two (*yama*)-seven (*adri*)-one (*indu*), and the digits are 17245. [439]   20

The planet's own orbit multiplied also by the quotient ⟨obtained⟩ from the ⟨number of⟩ terrestrial days ⟨in a *yuga*⟩ becomes the divisor of the product of the reduced orbit of the sky by the number of days ⟨elapsed in the *Kaliyuga*⟩[440]. Since the previously written quantity that has ⟨such⟩ remainders for two divisors is the product of the number of days ⟨elapsed in the *Kaliyuga*⟩ and the reduced orbit of sky, therefore, having divided ⟨it⟩ by their own divisors, the quotient is the passed revolutions of the sun and the moon. For the sun, 3723; for the Moon, 49777. For the two remainders ⟨obtained⟩ here ⟨by the two divisions⟩, the corresponding residue of revolutions have been indicated (in example 24). When that which has ⟨such⟩ remainders is divided by the reduced orbit of the sky, the quotient is the number of days ⟨elapsed in the *Kaliyuga*⟩,   six (*rasa*)-thirteen (*viśva*) multiplied   25 by the square of a hundred, and with digits also 1360000.

An example:

> **25. The two residues of minutes computed with the procedural compu-
> tation (*atividhikrameṇa*) called orbital, are respectively**
> **two-six (*aṅga*)-five (*iśu*)-four (*abdhi*)-five (*śilīmukha*, five arrows of love)-
> three-zero (*nabhas*)-five (*bhūta*)-five (*indriya*)-sixteen (*aṣṭi*) for the
> Sun|**
> **For the moon , four (*kṛta*)-six (*rasa*) eight (*vasu*)- three (*agni*)-twenty-**   p.153,
> **four (*sūkṣmakā*) multiplied by *ayuta* (ten thousand)**   line 1
> **Their revolutions, etc., the number of days ⟨elapsed in the *Kaliyuga*⟩
> and that which has ⟨such⟩ remainders for two ⟨divisors⟩ should be
> stated for both, with these two ⟨residues⟩ ||**

---

[438]This states in other words, the rule given in the *Mahābhāskarīya* (1.20) and exposed in the Appendix d. It is also explained in the supplement for this commentary of verse.

[439]The number of terrestrial days in a *yuga*, according to the rule given in the *Āryabhaṭīya*, as explained in Appendix d is 1577917500. $\frac{1577917500}{91500} = 17245$.

[440]This states the previously noted quotients:

$$\lambda_S = \frac{136335744x}{49797813966}$$

$$\lambda_C = \frac{136335744x}{3724920000}$$

	For the sun	16550354562
		49797813966
Setting down:	For the moon also	2438640000
		3724920000

In this case, because a pulverizer cannot bring about simultaneously ⟨all the values requested⟩, one by one, with a pulverizer, the residues of revolutions of the sun and the moon, should be reduced by their own divisors, ⟨and⟩ with that procedure "one should divide the divisor of the greater remainder by the divisor of the smaller remainder" (Ab.2.32), the computation of the number of days ⟨elapsed in the *Kaliyuga* is performed⟩. It is as follows: The divisor and the residue for the Sun reduced by six are :
$$\dfrac{2758392427}{8299635661} \; .$$

A pulverizer with both the reduced divisor and residue, is considered.[441] In this case, when one has multiplied the residue of degrees by sixty and divided by that very reduced divisor, the minutes ⟨of the mean longitude⟩ of the sun are obtained, and the residue of minutes has been obtained in excess (*atiricyate*). That has been written indeed. In this case, this is considered: "Will sixty multiplied by what ⟨is sought⟩, when decreased by the residue of minutes give an exact division for the reduced divisor?" In this way, the residue of degrees has been obtained . And that is 7377318041.

Or else: "will sixty multiplied by what ⟨is sought⟩, when decreased by one give an exact division for the divisor reduced by six?" When one has brought also a pulverizer with one as subtractive, in this way (*iti*), with that, a computation of the residue of degrees and a computation of the minutes ⟨are carried out⟩. The

---

[441] As noted in the previous example, we know that,

$$\lambda_S = \frac{136335744x}{49797813966},$$

where

$$\lambda_S = M_S + \frac{R_S}{12} + \frac{B_S}{12 \times 30} + \frac{L_S}{12 \times 30 \times 60} + \frac{S_S}{12 \times 30 \times 60 \times 49797813966},$$

when considered in terms of revolutions.

In the example given here $\frac{S_S}{49797813966}$ is known. The problem solved uses the following:

$$\frac{60 \times [\frac{L_S}{60} + \frac{S_S}{60 \times 49797813966}]}{49797813966} = L_S + \frac{S_S}{49797813966}.$$

If we call $x_B = \frac{L_S}{60} + \frac{S_S}{60 \times 49797813966}$ ("the residue of degrees"), we have, according to the values given in the problem:

$$L_S = \frac{60x_B - 16550354562}{49797813966}$$

Or after a reduction by six:

$$y_B = \frac{60(x_B/6) - 2758392427}{8299635661},$$

where $y_B = L_S$. As it is computed by Bhāskara in the following, $\frac{x_B}{6} = 7377318041$.

pulverizer with one as subtractive and the quotient are $\dfrac{8161308400}{(59)}$. 20

With that pulverizer the previously written residue of degrees has been obtained.[442] Hence, once more, a pulverizer with that residue of degrees and thirty is performed.[443] "Will thirty multiplied by what ⟨is sought⟩, when decreased by the residue of degrees give an exact division for the divisor reduced by six?" The residue of signs has been obtained. And that is 5502346520. In this way, once again, a pulverizer with that is performed. "Will twelve multiplied by what ⟨is sought⟩ when decreased by the residue of signs give an exact division of just that divisor reduced by six ?" The residue of revolutions has been obtained.[444] And 25 that is 3225074097. Since this is produced with the reduced divisor and dividend, this when multiplied by six becomes the residue of revolutions of the previously stated example (ex. 24); therefore that is indeed what was previously written. The pulverizer with one as subtractive and the quotient are $\dfrac{7607999356}{(11)}$. With the quotients of the divisions one by one the computations of the signs, degrees and minutes ⟨are carried out⟩.

In this way for the Moon also , when one has reduced, in due order, the quantities of the divisor and residue by a hundred increased by eight multiplied by an *ayuta* (ten thousand), the placement is $\dfrac{2258}{3449}$.   p.154, line 1

In due order a pulverizer with these ⟨is performed⟩ as before. And the residue of degrees has been obtained with sixty, and that is 2222^{445}. With one as subtractive

---

[442]The mean longitude of the sun is the same as in example 24, therefore the same residue of degrees, minutes etc is found.

[443]By the same reasonings as previously, the problem solved here is:

$$y_R = \frac{30(x_R/6) - 7377318041}{8299635661},$$

where 7377318041 is the value found previously for $x_B/6$.

[444]By the same reasonings as previously, the problem solved here is:

$$y_M = \frac{12(x_M/6) - 5502346520}{8299635661},$$

where 5502346520 is the value found previously for $X_r/6$.

[445]As before, since

$$\lambda_C = \frac{136335744x}{3724920000}$$

the following problem is solved by a pulverizer:

$$y_B = \frac{60x_B - 2438640000}{3724920000}.$$

Or after reduction by 180000:

$$y_B = \frac{60(x_B/180000) - 2258}{3449}.$$

The value found for $x_B/180000$ is 2222

also, the pulverizer and the quotient are $\frac{1782}{31}$. Again when one has performed
a pulverizer with the residue of degrees as subtractive and thirty, the residue of
signs has been obtained.[446] And that is 304. With one as subtractive also, the
pulverizer and the quotient are $\frac{115}{1}$. Hence, once more when one has performed
a pulverizer with the residue of signs as subtractive and twelve, the residue of
revolutions has been obtained, and that is 1175.[447] With one as subtractive also,
the pulverizer and the quotient of the division are $\frac{2012}{7}$.

The same   residue of revolutions brought here, when multiplied by precisely a
hundred increased by eight multiplied by an *ayuta*, becomes the residue of revolutions of the previously stated example (ex.24) that is, "for the moon, four (*kṛta*)-
six (*rasa*), etc., multiplied by an *ayuta*", just as written previously.

In this way, when one has known the residue of revolutions of the sun and the moon,
with that ⟨procedure⟩ "One should divide the divisor with a greater remainder
by the divisor with a smaller remainder", by proceeding as before the passed
revolutions and the number of days ⟨elapsed in the *Kaliyuga* are obtained⟩ just as
previously written.

Or else, that which is produced by that ⟨pulverizer⟩ procedure with the previous
amount of the residue of the revolutions,   multiplied by zero-zero-the cube of six
and divided by the divisor for its own orbit is the amount of the residue of minutes
⟨as given in ex. 24⟩[448]; hence this is considered: "Does [zero-zero]-the cube of six
multiplied by what is sought, when decreased by the residue of minutes mentioned
separately for the Sun and the moon, produce one by one an exact division by the
divisors told for their respective orbits?" When a pulverizer is performed in this
way, the elapsed revolutions for respectively the sun and the moon and the amounts
of their residue of revolutions are obtained. These revolutions and the amounts of
the residues of revolutions ⟨obtained here⟩ are just as previously written (in ex.
24).

---

[446] In other words, the following problem is considered:

$$y_R = \frac{30(x_R/180000) - 2222}{3449}.$$

[447] In other words, the following problem is considered:

$$y_M = \frac{12(x_M/180000) - 304}{3449}.$$

[448] In other words, this is the problem considered:

$$z = \frac{3724920000 \times 21600x - (8299635661 - 2758392427)}{49797813966}.$$

[A pulverizer with three remainders when an orbital operation ⟨is used⟩]

In the same way, a pulverizer with three remainders too is computed. It is as follows:

An example:

> **26. The recorded revolution residue for the Sun is zero (*gagana*)-three (*agni*)-two (*dasra*)-zero (*gagana*)-twelve (*sūrya*)-four (*abdhi*)-three (*rāma*)-five (*iṣu*)**
>
> **Three (*rāma*)- six (*aṅga*)-four (*abdhi*)-zero (*viyat*)-three (*kṛṣānu*, fire)-three (*dahana*, fire)|**   25
>
> **⟨The recorded revolution residue⟩ for the Moon is zero (*ambara*)-zero-four (*veda*)-zero (*gagana*)-three (*rāma*)-four (*abdhi*)-two (*dasra*)-two**
>
> **Nine (*randhra*)-seven (*adri*)-zero (*ambara*)-seven-five (*bhūta*)-two (*yamala*). The residue for Jupiter (*guru*) is said to be‖**
>
> **The quantity (*nicaya*) determined by zero (*vyoma*)-zero (*abhra*) -four (*abdhi*)- five (*śara*)-five (*artha*)-seven-seven (*giri*) eight (*vasu*)-nine (*aṅka*)-six-six**   p.155, line 1
>
> **Five (*bhūta*)-one (*indu*)-nine (*aṅka*)-six (*rasa*)-three (*agni*). This is ⟨obtained⟩ from the so-called orbital ⟨computation⟩ |**
>
> **The ⟨quantity corresponding to⟩ the three remainders, the number of days ⟨elapsed in the *Kaliyuga* and the passed⟩ revolutions should be stated according to the rule with the number of these**
>
> **If various pulverizers have been mastered ⟨by you⟩ according to the procedure told at Aśmaka‖**

	For the sun	330463534120230
		472332265467510
Setting down:	For the moon	25707922430400
		35330866200000	10
	For Jupiter	3691566987755400
		5602254071175000

In this case, the difference of these remainders [for the Sun and the moon] is zero (*vyoma*)-three (*agni*)-eight (*vasu*)-nine-eight-six-eleven (*rūdra*)-six (*rasa*)-five (*śara*)-five (*bhūta*)-seven (*adri*)-four (*kṛta*)-zero (*ambara*)-three (*agni*), and with digits also 304755611689830.

A reduction [of the remainders and the divisors] by zero-nine (*aṅka*)-five (*śara*)-two (*yama*)-eight (*vasu*)-two (*dasra*)-six (*rasa*)-nine-one (*indu*), with digits 196282590, ⟨is performed⟩. When reduced by that, ⟨the reduced divisor⟩ for the Sun is nine-   15
eight-three (*agni*)-six (*rasa*)-zero (*ambara*)-four (*abdhi*)-two (*yama*), and with digits also 2406389; for the moon ⟨the reduced divisor⟩ is zero-zero (*ambara*)-zero (*akāśa*)-zero (*viyat*)-eight-one (*indu*), and with digits 180000; the reduced difference of remainders is 1552637.

A pulverizer with these two reduced divisors and with the reduced difference of remainders ⟨is performed⟩; what is obtained is seven (*svara*)-six (*aṅga*)-seven (*adri*)-three (*rāma*)-six (*aṅga*)-six (*rasa*)-seven (*adri*)-six (*rasa*)-eight (*vasu*)-three (*loka*), and with digits also 3867663767. When this is divided by the divisor with the smaller remainder, what remains is seven (*svara*)-six (*aṅga*)-seven (*adri*)-three (*dahana*), 3767. This is multiplied by the reduced divisor with the greater remainder, and again [multiplied] by the reducer, that is zero-nine (*aṅka*)-five (*śara*)-two (*yama*)-eight (*vasu*)-two (*dasra*)-six (*rasa*)-nine-one (*indu*) (196282590), and increased by the greater remainder; what is produced is that which has ⟨such⟩ remainders for two divisors, zero-zero (*ambara*)-four (*udadhi*)-zero (*viyat*)-three (*agni*)-two (*yama*)-zero (*ākāśa*)-five (*śara*)-five (*śara*)-seven (*adri*)-zero-one (*indu*)-six (*rasa*)-zero (*ambara*)-six (*aṅga*)-nine (*aṅka*)-seven (*adri*)-seven (*svara*)-one (*indu*), and with digits also 17796061107550230400.

When a pulverizer of that which has ⟨such⟩ remainders for two divisors with the remainder of the third divisor, the product of the two divisors of the complete procedure is performed the divisor ⟨which is⟩ two (*yama*)-six (*rasa*)-one (*indu*)-seven (*muni*)[449]-five (*śara*)-two (*aśvin*)-six (*rasa*)-seven (*adri*)-four (*jaladhi*)-five (*śara*)-seven(*muni*)-one(*rūpa*)-three(*dahana*)-seven(*adri*)-zero-eight-zero (*ambara*)-nine -seven (*muni*)-eight (*vasu*)-six (*aṅga*)-sixteen (*aṣṭi*) ⟨multiplied by a million⟩, and the setting down with digits too is: 16687908073175476257[450]162000000.

Here, the division of the divisor with the greater remainder ⟨just obtained⟩ by the third divisor stated ⟨in this example is made⟩, there zero remains. Just that zero is the pulverizer. Therefore, the previously obtained ⟨quantity⟩ which has ⟨such⟩ remainders for two divisors is that ⟨quantity⟩ which has ⟨such⟩ remainders for three ⟨divisors⟩ which is just as previously written. Its (the quantity for three divisors) division by the quantity ⟨expressed in digits⟩ of the *yojanas* produced from the orbit of the sky, that is, by zero (*akaśa*)-four (*udadhi*)-eight (*vasu*)-one (*rūpa*)-three (*śikhin*)-five (*śara*)-four (*kṛta*)-fourteen (*manu*)-three (*loka*)-nine (*aṅka*)-twelve (*ravi*), 1293144531840 ⟨is made⟩. The result is the number of days ⟨elapsed in the *Kaliyuga*⟩: five (*śara*)-eight (*vasu*)-one (*rūpa*)-six (*aṅga*)-seven (*adri*)-thirteen (*viśva*) 1376185.

In this way, this pulverizer operation, when considering ⟨it⟩ well is immeasurable like the crossing of the waters of the great ocean. Therefore ⟨the commentary on pulverizers⟩ comes to an end.

---

[449]All manuscripts insert *svara* (7) here. So the wrong number should go back to an ancestor of all five manuscripts.

[450]According to all manuscripts a 7 is inserted here.

Index